MEASURING *The* UNIVERSE

THE HISTORICAL QUEST TO QUANTIFY SPACE

KITTY FERGUSON

headline

First published in 1999
by HEADLINE BOOK PUBLISHING

First published in paperback in 2000
by HEADLINE BOOK PUBLISHING

10 9 8 7 6 5 4

ISBN 0 7472 5699 3

Typeset by
Letterpart Limited, Reigate, Surrey

Printed and bound in Great Britain by
Clays Ltd, St Ives plc.

HEADLINE BOOK PUBLISHING
A division of the Hodder Headline Group
338 Euston Road
London NW1 3BH

www.headline.co.uk
www.hodderheadline.com

To my brother, David,
who made himself ill as a child and caused a family
crisis, worrying about the size of the universe

Contents

Acknowledgements

The author wishes to thank the following, who have read portions of the manuscript, answered her questions, supplied background material and information, and made suggestions and corrections. Without this help, *Measuring the Universe* could not have been written:

Judy Anderson, Boyd Edwards, Caitlin Ferguson, Yale Ferguson, Carlos Frenk, Wendy Freedman, Margaret Geller, Owen Gingerich, Stephen Hawking, Jill Knapp, Helen Langhorne, P. Susie Maloney, Robert Naeye, Saul Perlmutter, Barbara Quinn, Allan Sandage, Bill Sheehan, Patrick Thaddeus and David Vetter.

Credits for plate section photographs

Tilting at Windmills
1951

When I was nine years old, my father suggested one morning that he and my brother and I go out and measure the height of the windmill on my grandparents' farm. My brother and I agreed that was a fine idea.

How would we do it? Climb the windmill, of course . . . at least my father would. My brother and I wouldn't be allowed to try anything so dangerous as that. When my father reached the top, there would still be the problem of how to measure the height. We didn't have a measuring tape that long. Would he take a yardstick and mark off the yards on the windmill as he climbed? Maybe he would drop the end of a long rope from up there and cut it off, and we would stand clear while the cut-off piece fell, and then we would measure it. That must be the plan, for he'd said my brother and I would help him.

My brother suggested that my father wouldn't need to climb the windmill at all. We could throw something over the top, just clearing it. Yes, I interrupted, attach a rope to the thing we threw, a rope with inches and feet marked on it, and then pull back on it gently so it would catch on the top of the windmill, and see what the measurement was to the ground! No, no, said my brother, who was two years younger than I but already very mathematically minded, we would measure the curve the object followed through the air. Good thinking, said my father, but, practically speaking, more difficult than the original problem of measuring the height of the windmill.

I asked whether we might walk away from the windmill and measure how much smaller it looked as we got further away. More good thinking, said my father, but there was a better way.

He gave us a hint. He'd thought it was a wise idea to wait for a sunny day . . . and no one would have to climb the windmill or take a walk or risk wrecking the windmill with a bad throw . . . and the only tools we'd need would be a yardstick and our eyes and brains and a pencil and paper to do some calculations. And although at this latitude it would be possible to measure the windmill precisely at noon, it would be easier at another time of day.

Neither my brother nor I was clever enough to see where this was leading until my father said, 'The windmill does more than just pump water, you know. It casts a shadow, and so does a yardstick,' and then we began to understand how the trick could be done. We would stand the yardstick upright and measure its shadow. Then we would measure the windmill's shadow. If a shadow *this* long went with a three-foot stick, then a shadow *that* long went with a windmill of thus-and-so height. My brother and I didn't know how to make the comparison. My father taught us how and then pointed out that there was actually a more primitive way to find the answer. Wait for the time of day when the three-foot yardstick cast a three-foot shadow. At that moment the length of the windmill's shadow would be the same as the height of the windmill. We decided to use our newly acquired mathematics first, and then we checked our answer by sitting out in the Texas sun, watching the shadow of the yardstick creep along the ground.

That's how we measured the windmill, while above our heads the giant structure thrummed and creaked with the watery, metallic sounds windmills in central Texas made in those days, doing its work, turning and pumping, adjusting its angle to catch a stronger breeze, not paying any attention to the mental exertions of three little people below who had captured its shadow.

I was elated. It seemed we had outwitted the windmill without so much as touching it, and now we knew a wondrous secret: Not the height of the windmill, but how to find it out. None of us thought to ask: why are we doing this? There was no need whatsoever for any of us three to know the height of a windmill that wasn't even our own.

Measuring is one of the more practical uses for mathematics, but our ability and desire to measure isn't always wrapped up with the need to know useful answers. Going with numbers where we can't go in person – whether that's to the top of a windmill or to the origin and borders of the universe – has been and still is one of humankind's favourite intellectual adventures. At the end of the twentieth century it has outrun our practical requirements by billions of light years.

Compared with the adventure of finding them out, the actual measurements often seem dry as dust: The Sun is 149.5 million kilometres away (mean distance). The nearest star is 4.3 light years. The 'Local Group' of galaxies covers an area about 3 million light years in diameter. The distance to the edge of the observable universe is 8 to 15 billion light years. We shake our heads at how large these numbers are or admit their largeness makes them meaningless, remember them for a day or maybe long enough for a school exam . . . and then forget them. Science trivia.

Not trivial at all when you realize how hard-won these numbers are and what ingenuity it took and still takes to find them out. Can we even begin to imagine what it would be like if no one knew them? The night sky sparkles with pinpricks of light. Are these all the same distance from us? Suppose we didn't know. Suppose none of our contemporaries knew whether the Sun orbits the Earth, or vice versa, or even how large the Earth is. Suppose no one had guessed there are mathematical laws underlying the motion of the heavens. How would – how did – anyone begin to discover these numbers and these relationships without leaving the Earth? What made anyone even think it was possible to find out 'how far'? Without going there. Without climbing the windmill.

In the pages to come we'll take many steps back, forget we know the measurements or how to make them, and join our ancestors as they tease this information out of a sky full of stars. The laboratory isn't a neat, sterile room where carefully controlled experiments take place. Events in the heavens happen in their own good time and not before, and they are often not repeatable. We have learned to take what's on offer and make the best of it.

Our human point of view is sorely limited. Until recently we had no ground on which to stand and take our measurements, no possible viewing platform, other than here on Earth. In the twentieth century we've travelled to the Moon and looked back at our planet from space and sent probes out into the far reaches of our solar system. But by universal standards, by the standards of the distances we've learned to measure and still hope to measure, how pitifully close to home that is.

This book is a chronicle of how men and women over the course of two and a half millennia have built a ladder of measurement from our doorsteps to the borders of the known universe, and how the adventure has changed our ideas about the shape and nature of the universe and our place in it. It is not a history of all astronomy. There are fascinating discoveries, both in Western astronomy and in other cultures, that I have had to remind myself have no direct bearing on our knowledge of distances, size and shape. With regret, I have left them out, though the temptation to embark on long digressions from the main theme of the book has been almost irresistible.

We shall however broaden our focus in another direction to examine the context in which the discoveries have taken place, for this is a story inextricably bound up with the rest of social, political and intellectual history. One of our tasks will be to look for reasons why a particular discovery or measurement happened when and where it did. What was it about that time and place, that society, that mindset or intellectual milieu, the available technology, the chain of previous discovery, the way some random occurrences fell out . . . perhaps most interesting of all, what was it about a specific individual that precipitated this advance in knowledge?

The story of our wanting to know 'how far' – to make ridiculously out-of-reach measurements – must surely have begun before the beginning of recorded history. The known story of our success began some 2,200 years ago in north Africa near the mouth of the Nile with the measurement of the circumference of our own planet, long before anyone was able to circumnavigate it. The Hellenistic librarian Eratosthenes didn't need to know the circumference of the Earth. Nevertheless, he set about measuring it and he did it in a remarkably simple way. We now call Eratosthenes the father of the science of Earth measurement, 'geodesy'. The word has the sound of 'odyssey' in it.

We learn from Eratosthenes what I learned from my father . . . and we shall see it demonstrated repeatedly in this book: what can't be measured directly – what it is unthinkable that we should ever measure directly – *can* be measured in roundabout, inventive ways. Today we are attempting to determine the distance to the borders of the observable universe, far beyond any pinprick of light we see with the naked eye in the night sky, and to measure time back to the origin of the universe. The numbers are indeed too enormous to make sense to our little minds. Nevertheless our little minds have figured them out, one resourceful step at a time, each step building upon the last. It has been a history of astounding improvement in our technology, particularly in the twentieth century, but more than that, a history of raw ingenuity. It still is. Frontier inventiveness is not out of date.

With the benefit of hindsight we may be tempted to exclaim, 'Of course! Why . . . *I* could have thought of that! The windmill has a shadow. Of course!' for many of the methods we've devised to measure distances to out-of-reach places are simple enough for nearly everyone to understand – only a little more complicated than measuring my grandfather's windmill. But to figure these things out for the first time . . . how impossibly clever!

Kitty Ferguson,
January 1999

CHAPTER 1

A Sphere with a View
Third Century BC

The great mind, like the small, experiments with different alterna-
tives, works out their consequences for some distance, and thereupon
guesses (much like a chess player) that one move will generate richer
possibilities than the rest . . . It still remains to ask how the great
mind comes to guess better than another, and to make leaps that turn
out to lead further and deeper than yours or mine. We do not know.

Jacob Bronowski

Ask who Eratosthenes of Cyrene was, and unless you are
talking to someone who specializes in the minutiae of
Hellenistic culture, you are unlikely to hear that he was
a man who attempted to fix the dates of the major literary and
political events from the conquest of Troy until his own time in
the third century BC, that he composed a treatise about theatres
and theatrical apparatus and the works of the best-known comic
poets of the 'old comedy'; that he suggested a way of solving a
problem that had tantalized mathematicians for two centuries –
'duplicating a cube'; that he let his voice be heard on the subject
of moral philosophy and felt it essential to criticize those who
were 'popularizing' philosophy, accusing them of 'dressing it up

in the gaudy apparel of loose women'. It is true for Eratosthenes, as it is for many celebrated figures, that the strokes of genius for which he is revered were only a minuscule part of a lifetime of achievement, and not necessarily the part he judged most important.

Nothing on the list above won Eratosthenes his place in the history books. Two additional accomplishments did: the invention of 'the sieve of Eratosthenes' — a method for sifting through all the numbers to find which are prime numbers; and his remarkably accurate measurement of the circumference of the Earth.

Dismiss any thought that before Columbus no one knew the Earth was round. Admittedly, the shape of the Earth probably wasn't of much daily practical interest to most people in the ancient world. However, long before even Eratosthenes, those few who were wondering about it at all were not seriously suggesting that the Earth was flat or, indeed, any shape but spherical. The Pythagoreans, a school of thinkers with particular genius for mathematics and music, had decided as early as the sixth and fifth centuries BC that the Earth is a sphere. Plato, still a century before Eratosthenes, pictured a cosmos made up of spheres within spheres, nested one within another, with a spherical Earth at the centre. Aristotle, only a little later than Plato, vigorously subscribed to the idea of a spherical Earth, and his defence proved convincing not only to the ancient world but also to the Middle Ages. The idea that scholars of the Middle Ages believed the world was flat is, in fact, a myth created in modern times.

Aristotle used a number of arguments. During an eclipse of the Moon, the shadow cast by the Earth on the Moon is always curved. When we on the Earth move from north to south or vice versa we notice what appears to be a change in the position of the stars in relation to ourselves. In Aristotle's words:

There is much change, I mean, in the stars which are overhead, and the stars seen are different, as one moves northward or southward. Indeed there are some stars seen in Egypt and in the neighborhood of Cyprus which are not seen in the northerly regions; and stars which in the north are never beyond the range of observation, in those regions rise and set. All of which goes to show not only that the Earth is circular in shape, but also that it is a sphere of no great size: for otherwise the effect of so slight a change of place would not be so quickly apparent.

Aristotle speculated that the oceans of the extreme west and the extreme east of the known world might be 'one', and he reported with some sympathy the arguments of those who had noticed that elephants appeared in regions to the extreme east and the extreme west, and who thought therefore that those regions might be 'continuous'.

These reasons for belief in a spherical Earth came from observation, but Aristotle also argued on the basis of his philosophy. In that philosophy, five elements, earth, air, water, fire and aether, each have a natural place in the universe. The natural place for the element earth is at the centre of the universe, and for that reason earth (the element) has a natural tendency to move towards that centre, where it must inevitably arrange itself in a symmetrical fashion around the centre point, forming the sphere we call *the* Earth. Aristotle reported that mathematicians had estimated the Earth's circumference to be 400,000 stades; that is, about 39,000 miles or 63,000 kilometres (more than half again as large as the modern measurement). No record survives of the method they used to arrive at that number.

When Aristotle died in 322 BC at the age of 62, the military campaigns of his most highly achieving pupil, Alexander the Great, had just ended with the death of Alexander. Though there is a tendency to speak of 'the Greeks' and toss names like Eratosthenes into that file, the civilization and the culture we are dealing with after Alexander was much larger both in territory and concept than what is implied in the word 'Greek'. Alexander's campaigns had carried Greek knowledge, language and culture throughout Asia Minor and Mesopotamia as far east as present-day Afghanistan and Pakistan, all the way to the Indus River, as well as to Palestine and Egypt. Vastly widened intellectual horizons were part of his extraordinary legacy. The culture of Greece and its colonies and the cultures of the conquered peoples began to mix and marvellously enrich one another. This was the dawn of the Hellen*istic* era, as opposed to the Hellenic era. That is, Greek*ish*, as opposed to Greek.

At the time of Alexander's and Aristotle's deaths, within a year of one another, Athens was still the undisputed centre of the intellectual world. That pre-eminence was not to last. Alexander's generals divided his empire, and Ptolemy's portion was Egypt and Palestine. He made Alexandria, near the mouth of the Nile, his capital. This already prospering city began to grow in size and splendour, and Ptolemy and his successors, reputedly ruthless in their exploitation of the lands under their control, amassed a surplus of wealth, some of which they chose to spend on literature, the arts, mathematics and science. Scholars are divided as to which Ptolemy should get the credit (Ptolemy's successors were also named Ptolemy), but either the first or the second of them, and perhaps it took both, decided to extend the royal patronage to found a library and museum. This was not an institution of the sort we call 'museum' today. As that name suggests, it was a temple to the muses, which meant both a religious shrine and a centre of learning.

Meanwhile the old, justly famous schools across the sea in Athens – schools founded by Plato, Aristotle, Epicureus and the Stoics – were no longer producing vibrant new ideas to quite the extent they had once done, though they were still the places a young man of Eratosthenes's time would have wished to go for his education. Alexandria began to rival and eventually supplanted Athens as the focal point of the intellectual world, and the museum and library there became *the* premier research institution. The library grew large, containing by one ancient estimate nearly 500,000 rolls. Eratosthenes was its director or librarian at the end of the third century BC, with a salary provided from the royal coffers.

Most of us have heard that one of the devastating tragedies in the history of humankind was the burning of the contents of the library at Alexandria. The story (now thought to be apocryphal) is that all those rolls were burned to heat the public baths for six months in the seventh century AD. Today there is a campaign underway to raise funds to rebuild the structure, but that effort seems rather pitiful and beside the point, in view of what can never be retrieved – the assembled knowledge of our ancestors in antiquity, hard-won over many centuries. We sense that something dreadful happened to us with the loss of all those rolls, whether it occurred quickly and calamitously in the seventh century or, more likely, gradually through neglect and the many political, military and religious turns of fortune that affected the city of Alexandria prior to that. Perhaps its loss was the symbol and symptom of a greater tragedy: the increasing lack of any widespread perception that such intellectual achievement was valuable. By the seventh century, there was probably little left to burn. It took centuries for humanity in the Western world to reach again an intellectual level on a par with the civilization that had produced that lost collection. But when Eratosthenes was librarian (235–195 BC) that was all in the future. He knew the Alexandria library in its heyday.

Scholars in the Hellenic and Hellenistic worlds would have been mystified by our present-day concept of 'science' as a distinct category of knowledge and pursuit of knowledge. They had several different words for what we call 'science'. Some modern words have evolved from these terms, but the modern words don't have precisely the same meaning these had in ancient Athens and Alexandria. Some examples are: *peri physeos historia* (inquiry having to do with nature); *philosophia* (love of wisdom, philosophy); *theoria* (speculation); and *episteme* (knowledge). Hellenistic scholars thought of 'physics' as one of three branches of philosophy. The other branches were 'logic' and 'ethics'.

The financial support of the Ptolemys and their efforts to outbid all competitors when it came to collecting the masterpieces of Greek literature and encouraging distinguished scholars to flock to Alexandria were motivated by desire for prestige – to add to the lustre and apparent power of the dynasty. They were also far from displeased when research could be applied to problems connected with weaponry. However, a key difference between the ancient way of thinking and ours is that although Hellenic and Hellenistic scholars didn't ignore the possibility that their study might serve practical purposes, they were much more inclined to justify their work on the grounds that it contributed to wisdom, or improved one's character, or led to a greater appreciation of the beauty of the universe and understanding of its creator. It seemed not to occur to these men and women that their efforts might hold the key to material progress. The work was its own reward, an end in itself, not a means to an end. The life of a scholar, the life of 'contemplation', was considered to be an exquisitely happy one. Doctors, whose efforts were intended to have more everyday practical value, were apt to differentiate themselves entirely from the 'philosophers'.

Related to this mindset that sees intellectual exercise as an end in itself is a perspective in which *how* to solve a problem is equally as interesting as actually solving it, often more so. This attitude arose partly out of necessity, for Greek and Hellenistic scholars were avidly interested in questions that they lacked the technology to answer definitively. Perhaps we can best acclimatize ourselves to the ancient way of thinking by recalling doing mathematics in school. Presented with the problem 'If you ride your bicycle at an average of 30 miles per hour, and it takes you 10 minutes to get to school, how far is school?' you do not immediately start quibbling that 30 miles per hour is not an accurate measurement of the speed you normally ride, that it actually takes you 12 minutes to get to school, and that this exercise isn't going to end with anyone knowing how far your school really is. No. What everyone is interested in is your showing that you understand how to solve the problem. Move back a step and imagine that it was also up to you to invent the method for solving it – that no one, in fact, had ever even thought it possible to calculate the distance to your school, and that you couldn't ride there to measure it directly – and you have put yourself a little in the shoes of Eratosthenes and other scholars of the Hellenic and Hellenistic world. It is an attitude which allows, indeed encourages, the formation of hypotheses, sometimes out of thin air, statements such as 'We don't know that this is true, but let's assume for a moment that it is, and see where that gets us.' Or even such a statement as 'We know that this is *not* true, but let's pretend for the moment that it is and ask "what then?" ' To criticize the results of an exercise like that by saying the results are 'wrong' (i.e., do not accord with twentieth-century findings) is to miss the point.

Does this mindset in which the pursuit of knowledge was valued quite apart from any practical spin-off – where the method was often thought more important than the result, where hypotheses, even those based on false assumptions, were encouraged – provide a clue to Eratosthenes's success? Perhaps. After all, the Hellenistic world had no practical need to know the circumference of the Earth. However, that attitude pre-dated Eratosthenes by several centuries. What's more, Eratosthenes's results were remarkably *accurate* by twentieth-century standards. Did his success have something to do with the widening of mental horizons and the mixture of knowledge from many cultures that followed the campaigns of Alexander? It did, but, again, Eratosthenes was not the only inventive man to enjoy that legacy. And the notion that the Earth isn't flat was nothing new either, nor was the idea of calculating its circumference. There had been estimates in Aristotle's time. But it was Eratosthenes who did the measurement and got it right – or so close to right that his calculation impresses us to this day – using a line of thought that we can easily agree was correct.

Eratosthenes, 'son of Aglaos', was born not in Egypt or Greece but in the ancient city of Cyrene, west of Egypt on the northern coast of Africa. Citizens of Crete and Santorini had founded Cyrene some 350 years earlier and it had become one of the most cultured cities of the Hellenistic world, though still subordinate to the Egypt of the Ptolemys. Cyrene counted some distinguished figures among its citizenry. Besides Eratosthenes there was Aristippos, who founded the Cyrenaic school there. He was a pupil of Socrates. Aristippos's daughter Arete followed him as head of the school. Her son Aristippos II succeeded her. He was nicknamed Metrodidactos, which translates as 'mother-taught'.

The date given for Eratosthenes's birth is the '126th Olympiad', referring to the Olympic games that took place every four years. In modern dating, that puts it between 276 and 273 BC. He received most of his education in Athens at the feet of eminent scholars of the New Academy and the Lyceum. Plato and Aristotle had originally founded these schools (in Plato's day the New Academy had been simply the Academy) and, though much had changed about them by the time Eratosthenes arrived, one still couldn't do better by way of an education.

By the middle of the century Eratosthenes had written a few philosophical and literary works and some of these had come to the attention of Ptolemy III Euergetes. The 'brain-drain' from Athens being in the general direction of Alexandria, Eratosthenes in about 244 BC agreed to move there and become a fellow of the museum and tutor to the prince, Philopator. (It isn't to Eratosthenes's credit that his pupil, though a patron of arts and learning, gained a reputation for dissipation and crimes equal to Nero's and Caligula's later in Rome.)

In the course of time Eratosthenes became a senior (alpha) fellow of the museum and upon the death of the chief librarian took over that post – an absolutely unparalleled vantage point from which to keep up with everything that was going on in the intellectual world.

Eratosthenes's colleagues gave him two nicknames: *pentathlos* and *beta*. The word *pentathlos* came from athletics. It was a name for those who entered the 'pentathlon', which required five skills: jumping, discus throwing, running, wrestling and either boxing or javelin throwing. Eratosthenes was no athlete. The nickname for him implied a jack-of-all-trades. *Beta* means 'B' or number two, or second. Put those together and you get 'jack-of-all-trades and master of none'. Whether these names were fondly or snidely given isn't clear. Probably snidely. It seems a bit odd that in an era following so closely on the heels of Plato and Aristotle, who had their fingers in just about every intellectual pie around, a man should be mocked for being a jack-of-all-trades. Perhaps in the three-quarters of a century between Aristotle's death and Eratosthenes's arrival at Alexandria, scholarship had become more specialized and specialists had begun to sneer at those who were not specialists. Eratosthenes was evidently old-fashioned to be such a polymath, but he had been educated that way . . . and how could a man focus very narrowly when he was head of the library, *the* repository of knowledge and ideas on every subject, holding a job which made him responsible for helping the Ptolemys add to that collection. Eratosthenes was bombarded daily by new thoughts and discoveries. Modern scholars compare the breadth of his knowledge to that of Aristotle and Leonardo da Vinci. Whatever criticism Eratosthenes endured, his eclecticism served him well. 'Beta' is remembered while some who dubbed him that have long since been forgotten.

Unfortunately, none of Eratosthenes's many works have

survived except in fragments. It's not even certain that all the fragments attributed to him are genuine. Most information about him comes through reports and references of others. However, there is enough to tell that Eratosthenes's measurement of the Earth, and also his motive for attempting it, were rooted in his eclectic and far-ranging knowledge and interests. Eratosthenes was a man of the world, in the real sense of those words. He refused to categorize people as Greeks opposed to Barbarians, adopting a more cosmopolitan attitude which differed from the mindset of Greece in earlier centuries but was not uncommon in Hellenistic times. Perhaps it isn't inaccurate to use a modern term and call this a global point of view. Eratosthenes not only thought that way, but he followed through by collecting information about the people, products and geography of far-flung areas. He wrote about the history of geographical measurement, recalling old ideas going back to Homer about the size, shape and geographical lay-out of the Earth. In fact, he did nothing less than pull together virtually all the geographical knowledge that had been accumulating up until his own time.

Over the centuries, this material had taken a variety of forms. It came from traders, explorers, travellers – as well as mathematicians and philosophers – and it ranged from fantastic tales to more straightforward reporting, from speculation to measurements and estimates resting on what were probably recognized as shaky assumptions. Among the more reliable sources were eye-witness accounts of Alexander the Great's expeditions and the measurements and records of distances covered on those marches. There were itineraries of coastal voyages and maps and charts connected with them. There was a treatise on harbours by Timosthenes, the admiral of the Ptolemaic fleet, who also studied the winds. There was a book entitled *On the Ocean* by the merchant sea-captain Pytheas, who in about 320 BC sailed north along the coast of Spain and France and reached Cornwall, then continued all the way up to the Orkneys and the Shetlands to latitudes near those of the midnight sun. Pytheas took bearings throughout his voyage and recorded them in his book, which also has descriptive passages:

> The barbarians showed us where the Sun keeps watch at night, for around these parts the night is exceedingly short, sometimes two and sometimes three hours, so that only a short interval passes after the Sun sets before it rises once more.

Eratosthenes respected Pytheas's information, though it must have seemed almost as fantastic as Homer's, while many other scholars were contemptuous and disbelieving. Living as Eratosthenes did in Hellenistic Egypt, he may also have known of centuries-old and astoundingly accurate Egyptian geographical calculations.

Eratosthenes's expertise on longitude and latitude surpassed any other of his day or earlier. His predecessors had divided the map into zones. He took that work several steps further by improving on a map devised about 25 years before his birth by a man named Dicaearchus of Messene. Dicaearchus had divided the known world by using two lines or bands that intersected one another – one running east–west, the other north–south. On Eratosthenes's revised map the two lines crossed at Rhodes, a little to the east of where Dicaearchus's lines had met. The horizontal line passed near Gibraltar (then known as the Pillars of Hercules), ran the length of the Mediterranean and then followed the Taurus chain of mountains in southern Turkey (Toros Daglari on modern maps). That line is remarkably near to following what we call the 37th parallel – an impressive achievement for men without the benefit of the mathematical and astronomical knowledge that would go into later mapmaking. It was not yet possible to figure out latitudes with very great precision, and it was virtually impossible to determine longitude (which would prove to be a problem in Eratosthenes's measurement of the Earth). On Eratosthenes's map the vertical line followed the Nile, which doesn't line up so perfectly with Rhodes on modern maps. He added six further lines drawn vertically at intervals between the western and eastern boundaries of the inhabited world, and six more horizontal lines drawn at intervals between its northern and southern boundaries, and he established and measured geographic zones, dividing the world horizontally between the tropical region, the temperate region and the polar circles.

Eratosthenes was also well-acquainted with state-of-the-art geometry, both from Euclid's profound summing-up about 25 years before Eratosthenes's birth and from his association with Archimedes, one of the towering creative geniuses that Greek and Hellenistic civilization produced and also one of history's greatest eccentrics. Most schoolchildren have heard the tale of Archimedes solving a mathematical problem in his bath, leaping from the water, and running naked through the streets shouting 'Eureka!' This avid mathematician eventually lost his life when Roman troops sacked Syracuse. Archimedes, so the story goes, was drawing a mathematical figure in the sand when a Roman soldier (who had missed hearing an order from his superiors to respect the person of this famous old man) asked him to pack up and move along. Archimedes unwisely told the soldier not to interrupt his thought process.

The Hellenistic world revered Archimedes as an inventor (though he himself dismissed such practical achievements as unworthy of notice) and a useful man to have around in a war. Legend has it that he destroyed an entire Roman fleet by using burning mirrors. The Middle Ages thought of him as an engineer and a wizard and credited him with the invention of the Staff of Archimedes. This device was a stick with a small flat disc that could be run up and down it. An observer held the stick up to the Sun and moved the disc along it until it appeared to cover the Sun, then noted on a scale the distance from disc to eye, thus deriving the Sun's apparent diameter.

Modern history and mathematics books recall Archimedes as a brilliant mathematician and geometer who contributed significantly to the understanding of circles and spheres. Archimedes was in the habit of sharing his discoveries and his methods with Eratosthenes and even dedicated his greatest work, *Method*, to him. Eratosthenes must have welcomed another scholar who was almost as eclectic as he was himself.

Eratosthenes's thoughts stretched to the horizon in all directions. Perhaps it follows that he would have longed to know not only what was beyond those horizons but how far 'beyond' was? Mapping and systematizing things geographically was his bent. Would he not have been unusually curious about how large the total map was? How remarkable if it really should turn out to be, as Aristotle speculated, 'a sphere of no great size'! Eratosthenes's thoughts often took a historical turn, and he was aware of previous attempts to measure the Earth or estimate that measurement. Would he not have wanted to try his own hand at it, using Euclid's and Archimedes's newer understanding of geometry?

There is still one circumstance to be mentioned – a simple, trivial matter, yet Eratosthenes's successful measurement of the circumference of the Earth would not have taken place without it. A happenstance, perhaps, that such a small gem of information reached the ears of this man who realized what it meant and what could be done with it. It is true that the fact that this snippet of news reached him *did* have something to do with the broadened mental horizons of the world, with improved communications from remote areas, with Eratosthenes's own world centring on northern Africa, and with his habit of keeping his ears and eyes open and wanting to know everything and anything. He was indeed the right man in the right time and place. Perhaps there was no other so likely to run across this back-page news and recognize its worth:

In a well located at Syene (near modern Aswan), on the day

of the summer solstice, a shaft of sunlight penetrated all the way to the bottom of the well.

To Eratosthenes there was nothing trivial about this information. It meant that the Sun was shining directly down at Syene, not at an angle, and he knew this showed that Syene was on the tropic. A stick set up at noon at Syene on the day of the summer solstice would not cast a shadow. A stick set up at Alexandria (which he thought was the same longitude as Syene) *would*. Accordingly, Eratosthenes set up a stick at Alexandria on the day of the summer solstice and measured the angle of its shadow when that shadow was at its shortest.

Figure 1.1 below shows the stick at Alexandria and its shadow, and what is meant by 'the angle of the shadow'. The figure illustrates how that must also be the angle 'subtended' by the arc Syene–Alexandria at the centre of the Earth. To put that

Figure 1.1

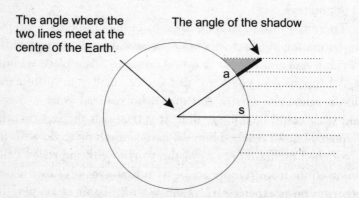

Because the Sun's rays are running parallel as they strike the Earth, if a line is drawn from Alexandria (a) where the stick casts a shadow, to the centre of the Earth, and a second line from Syene (s) where there is no shadow, to the centre of the Earth, the angle where those two lines meet will be the same as the angle of the shadow at Alexandria.

a bit more simply: if we draw a straight line from the point marked Alexandria (where the stick is casting a shadow) to the centre of the Earth, and a second straight line from the point marked Syene (where the stick casts no shadow) to the centre of the Earth, those lines will of course meet at the centre of the Earth. We want to know the angle between those two lines where they meet. Geometry tells us, as it told Eratosthenes, that the angle at the centre of the Earth and the angle of the shadow at Alexandria will be the same angle.

Figure 1.2 illustrates Eratosthenes's measurement.

Figure 1.2

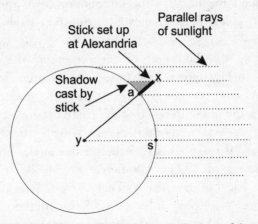

Because the sunlight shone all the way to the bottom of the well at Syene (s), Eratosthenes knew that the Sun was shining straight down on the Earth there. He set up a stick at Alexandria (a), where the Sun wasn't shining straight down, and he measured the angle (x) of the shadow cast by the stick. He knew that because the Sun's rays all run parallel as they strike the Earth, the angle (y) where a line drawn straight down from Alexandria and a line drawn straight down from Syene would meet at the centre of the Earth would be the same angle as the angle of the shadow cast by the stick (x). If Syene is due south of Alexandria, then the distance between Syene and Alexandria must be the same fraction of the Earth's total circumference as the angle at x or y is of 360°

15

Eratosthenes found that the shadow angle at Alexandria was 7⅕°, and so he knew that the angle between the 'Syene–Alexandria lines' (meeting at the centre of the Earth) was also 7⅕°. A circle has 360°, and it is a simple process to find out how many of the Syene–Alexandria angles (7⅕°) it will take to make 360°. Think of the cross-section of the Earth as a pie and the two lines coming from Syene and Alexandria as cutting out a wedge of pie. How many wedges of that size can you cut from the whole pie? Divide 360 by 7⅕, and it comes out to 50 wedges. If we say (as Eratosthenes did) that the distance between Syene and Alexandria at the surface of the Earth (at the pie-crust edge of the pie) is '5,000 stades', then we can multiply 5,000 by 50 and conclude that the distance all the way around the Earth – the circumference of the Earth – is 250,000 stades. Eratosthenes later fine-tuned this to 252,000 stades.

What is this odd unit of measurement, the stade? That question brings up a problem in evaluating Eratosthenes's result. Whether or not that result matches modern measurements for the circumference of the Earth depends on the length of 'stade' he was using, and it isn't known exactly what that length was. If there are 157.5 metres in a stade, Eratosthenes's result comes to 39,690 kilometres or 24,608 miles for the circumference of the Earth. That is very near the modern calculation – 24,857 miles (40,009 kilometres) around the poles and 24,900 miles (40,079 kilometres) around the equator. After he had found the circumference, Eratosthenes calculated the diameter of the Earth as 7,850 miles (12,631 kilometres), close to today's mean value of 7,918 miles (12,740 kilometres).

Another way of figuring a stade was as ⅛ or ⅒ of a Roman mile, and that would make Eratosthenes's result too large by modern standards. There was one additional small difficulty. Eratosthenes assumed that Syene lay on the same line of longitude as Alexandria. Actually, it does not.

But this is nit-picking! No apology need be made for Eratosthenes. First of all, he arguably came astonishingly near to matching the modern measurement. Second, he was probably, for all his curiosity about the world, enough a man of his time to find the puzzle of how to solve this problem by the imaginative use of geometry at least as interesting as the actual measurement. The *method* is ingenious and it is correct. If the numerical result is a little fuzzy because of a lack of agreement about the length of a stade and the impossibility of determining longitude precisely, that does not prevent our recognizing what a brilliant achievement this was or appreciating the intellectual leap involved in recognizing that it *could* be done and *how* it could be done.

Eratosthenes didn't focus his thoughts only on the Earth. He also raised them above the horizon to consider astronomical questions of his day. When it came to measuring the distances to the Sun and the Moon, he must have realized that he had no tool at his fingertips to equal the news about the well in Syene. Nevertheless, he gave it a try, with far less success than he had in measuring the Earth's circumference.

Another Hellenistic scholar, Aristarchus of Samos, also tried to measure the distances to the Moon and Sun. Little information exists about him as a person. He lived from about 310 to 230 BC and would already have been a grown man when Eratosthenes was born. The island of Samos was under the rule of the Ptolemys during Aristarchus's lifetime and it is possible that he worked in Alexandria. Archimedes was certainly aware of his contributions.

The only written work of Aristarchus that has survived is a little book called *On the Dimensions and Distances of the Sun and Moon*. In it he describes the way he went about trying to determine these dimensions and distances and the results he got.

The book begins with six 'hypotheses':

1.   The Moon receives its light from the Sun.

2. The Moon's movement describes a sphere and the Earth is at the central point of that sphere.
3. At the time of 'half Moon', the great circle that divides the dark portion of the Moon from the bright portion is in the direction of our eye. (In other words, we are viewing the shadow edge-on.)
4. At the time of 'half Moon', the angle (at the Earth as shown in Figure 1.3) is 87°.
5. The breadth of the Earth's shadow (at the distance where the Moon passes through it during an eclipse of the Moon) is the breadth of two Moons.
6. The portion of the sky that the Moon covers at any one time is equal to 1/15 of a sign of the zodiac.

Aristarchus's fourth and sixth assumptions are both far from accurate. The actual angle at the Earth in Aristarchus's triangle would be 89° 52', not 87°, and 89° 52' is very close to 90°. The

Figure 1.3

Aristarchus's measurement of the relative distances to the Moon and the Sun: when the Moon is a half Moon, the angle at the Moon (in this triangle) must be 90°. So a measurement of the angle at the Earth determines the ratio of the Earth–Moon line to the Earth–Sun line; in other words, the ratio of the Moon's distance to the Sun's distance.

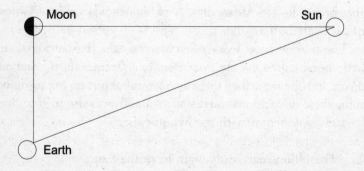

angle at the Moon in Aristarchus's triangle *is* 90°. That makes lines B and C so close to parallel that, on a drawing, the triangle would close up and be no triangle at all. The portion of one sign of the zodiac that the Moon covers is not $\frac{1}{15}$, and it isn't clear why Aristarchus, who must have known this from observation, chose that value.

Aristarchus's results are not what we now measure these relative distances to be. By his calculation, the distance to the Sun is about 19 times the distance to the Moon and the Sun is 19 times as large as the Moon. The modern ratio between their distances is 400 to one. The measurement Aristarchus was trying to make was extremely difficult with the instruments available to him. It is no simple undertaking to determine the precise centres of the Sun and the Moon or to know when the Moon is exactly a half Moon. Aristarchus chose the smallest angle that would accord with his observations, perhaps to keep the ratio believable. Throughout antiquity and the Middle Ages, estimates of the relative distances to the Sun and Moon would continue to be too small.

Aristarchus didn't stop with estimating the ratios, but found ways of converting them into actual numerical distances to the Sun and Moon and diameters for both bodies. He could see that the *apparent* size of the Moon and the Sun (meaning the size they appear to be when viewed from Earth) are about the same. During a solar eclipse, the Moon just about exactly covers the Sun. To put that in more technical language: they both have approximately the same 'angular size'. Angular size tells how much of the sky a body 'covers' and is measured in 'degrees of arc'. Both the Sun and the Moon have angular sizes of about one half of a 'degree of arc'. (For a fuller explanation of those terms, see Figure 4.5.) For that to be true, the two bodies don't actually have to be the same size, for how large they appear when viewed from Earth also depends on how distant they are. (See Figure 1.4a.) Aristarchus assumed that the Sun is much larger than the Earth, and that it was safe to assume also that the

Figure 1.4 Aristarchus's calculation of the size and distance of the Moon.

a.

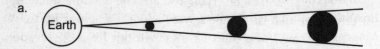

Surprisingly, all three of these bodies look the same size when viewed from Earth. We observe the 'angular size' of a body like the Moon or Sun, not its true size. It could be small and close or large and far away and still have the same 'angular size'. Aristarchus saw that the Moon and the Sun have about the same angular size; that is, they *look* the same size when viewed from Earth, but he knew they are not the same true size.

b. (The angles shown in this drawing are much larger than those that really exist.)

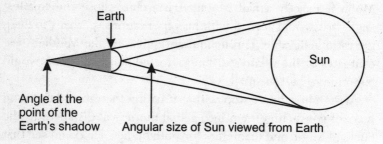

Aristarchus assumed that the Sun is much larger than the Earth. If that is true, then the angle at the point of the Earth's shadow is about equal to the angular size of the Sun as viewed from Earth.

c.

Observing an eclipse, Aristarchus concluded that the breadth of the Earth's shadow where the Moon crossed it was approximately twice the diameter of the Moon. He knew the angle formed at the point of the Earth's shadow and also the angular size of the Moon. There was only one distance to put the Moon where it would cover half the area of the shadow.

Note: These drawings are not to scale.

shadow cast by the Earth has about the same angular size as the Sun and the Moon (½ a degree of arc). (See Figure 1.4b.)

Aristarchus arrived at his fifth 'hypothesis' above – the breadth of the Earth's shadow (at the distance where the Moon passes through it during an eclipse of the Moon) is the breadth of two Moons – by observing a lunar eclipse of maximum duration, which means an eclipse in which the Moon passes through the exact centre of the Earth's shadow. He measured the time that elapsed between the instant that the Moon first touched the edge of the Earth's shadow and the instant that it was totally hidden. He then found that that length of time was the same as the length of time during which the Moon was totally hidden. He reasoned that the breadth of the Earth's shadow where it was crossed by the Moon must therefore be approximately twice the diameter of the Moon itself (Figure 1.4c). If, as he thought, the angle formed at the point of the Earth's shadow was the same as the angular size of the Moon, that gave him only one distance at which to put the Moon where it would cover half of the area of the shadow.

Aristarchus concluded that the Moon was ¼ the size of the Earth, and that the distance to the Moon was about 60 times the radius of the Earth. Both of those values are close to the modern values. Using Eratosthenes's calculation of the Earth's radius, Aristarchus arrived at an actual distance to the Moon in stades. He had less success with the distance to the Sun. His earlier estimate – that the Sun's distance is about 19 times the Moon's distance – was in error, and a second approach he tried, though it was ingenious and correct, required timing the phases of the Moon with a precision impossible in his day.

It was another of Aristarchus's ideas that secured his place much more firmly in the annals of astronomy. Hearing of it, one has a chilling sensation of stumbling into a prophetic vision. For Aristarchus suggested, 17 centuries before Copernicus, that the Earth is not the unmoving centre of everything but instead moves round the Sun, and that the universe is many times larger than anyone in his time thought – perhaps infinitely large.

For centuries it had been widely assumed that the Earth was the centre of everything. The accepted picture of the cosmos was a series of concentric spheres – spheres embedded one within the other – with the Earth resting motionless at the centre of the system. (See Figure 1.5.)

Plato and Euxodus of Cnidus, a younger contemporary of Plato, had introduced this model, and Aristotle's model of the universe was a further development of it, though he differed from Euxodus as to the number and nature of the spheres. However, it wouldn't be correct to think that everyone, without

Figure 1.5

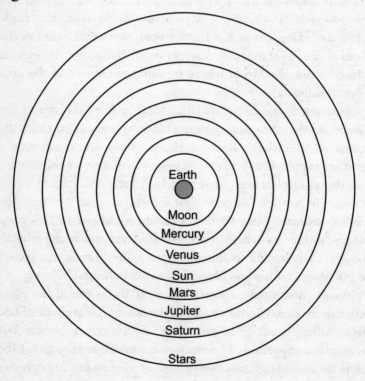

exception, since the dawn of human thought had agreed that the Earth was the centre and didn't move. Some Pythagorean thinkers had decided in the fifth century BC, largely for symbolic and religious reasons, that the Earth was a planet and that the centre of the universe must be an invisible fire. Heraclides of Pontus, a member of Plato's Academy under Plato, proposed that the daily rising and setting of all the celestial bodies could be nicely explained if the Earth rotated on its axis once every 24 hours.

But Aristarchus went further. Although information about his theory of a Sun-centred cosmos comes second-hand, no one disputes his authorship of the idea because there is plenty of secondary evidence. According to Archimedes:

> Aristarchus of Samos brought out a book of certain hypotheses, in which it follows from what is assumed that the universe is many times greater than that now so called. He hypothesizes that the fixed stars and the Sun remain unmoved; that the Earth is borne round the Sun on the circumference of a circle . . . and that the sphere of the fixed stars, situated about [that is, centred on] the same centre as the Sun, is so great that the circle in which he hypothesizes that the Earth revolves bears such a proportion to the distance of the fixed stars as the centre of the sphere does to its surface.

Aristarchus had done no less than move the centre of the cosmos to the Sun. In this astounding turn-about, the Earth moves round the Sun and, rather than the sphere of the fixed stars making a revolution of the heavens once every 24 hours, it is the Earth that turns, rotating on its axis – as Heraclides had suggested. The stars are extremely far away. The implication is, infinitely far.

Did Aristarchus also speculate that the other planets move round the Sun? It would seem a logical next step, but there is no historical evidence that he did. In any case, it's unlikely that he understood the enormous significance of his model, that it provides, at a sweep, the basis for explaining the planets' positions and movements far more simply than a model with the Earth as centre. It is impossible to tell from the surviving evidence whether Aristarchus really was personally disposed to thinking that the Earth moved around the Sun or whether he made the suggestion merely for the sake of argument, as in 'Let's just suppose for the moment that this is how things work.' Why did this revolutionary suggestion came at this time and place in history? The simple answer may be that this was an intellectual environment that encouraged one to make suggestions and put forward hypotheses, even hypotheses based on assumptions that were known to be incorrect – in no way claiming they were true – as the starting point for an interesting line of enquiry.

With Aristarchus the question must also be turned on its head to ask not only why this idea emerged but why it died at birth. Seleucus of Seleucia, a Chaldaean or Babylonian astronomer (Seleucia was on the Tigris river) in the second century BC, took Aristarchus's suggestion seriously – not merely as a hypothesis. Seleucus believed Aristarchus was right. However, no one else for the remainder of antiquity did, and the remainder of antiquity was by no means a dark age when it came to astronomy. It's only partly correct to blame this resistance on an ideological attachment to having Earth and humanity the centre of everything.

If surviving information can be trusted, public opinion reacted almost not at all. Aristarchus's idea must have been too far removed from common knowledge and common sense to draw much popular attention. The historian Plutarch reports one comment from the Stoic Cleanthes (the Stoics were reputedly weak in natural science and even 'anti-scientific') that Aristarchus of Samos ought to be indicted on a charge of impiety for putting the 'Hearth of the Universe' in motion.

There is no record of anyone trying to take Cleanthes's advice. Some philosophers scolded Aristarchus for trespassing in an area of knowledge that was *their* sole domain; and there were also complaints accusing him of undermining the art of divination.

As for astronomers, what mattered most to them was that there was no observational evidence whatsoever to support the vast distances to the stars that Aristarchus's scheme required, while there *was* observational and physical evidence that made his Sun-centred arrangement seem highly unlikely:

1. If the Earth moves round the Sun, then we on the Earth should see some variation in the positions of the stars, relative to one another, as we view them from different points along the Earth's orbit. No such variation had been observed (nor could it be with the technology available at the time). Aristarchus saw that this objection wouldn't be valid if the stars were far enough away. Hence his suggestion that they were very far away indeed, perhaps even at infinite distance. The fact that the position of the stars, relative to one another, does change as Earth orbits – that there is 'stellar parallax motion' – wasn't confirmed by observation until the mid-19th century.

2. If the Earth rotates on its axis, in fact, if it moves at all, this should have some noticeable effect on the way objects move through the air. Ancient astronomers realized that if the Earth rotates on its axis once every twenty-four hours, then the speed at which any point on its surface is moving is very great indeed. So, how could clouds, or things thrown through the air, overcome this motion? How could they ever move *east*? Even if not only the Earth but also the surrounding air rotates on the Earth's axis, solid bodies moving through the air should still in some way show the influence of the Earth's rotation.

3. It's plain to see that heavy objects travel towards the centre

of the Earth. If this law applies to heavy objects everywhere, then the centre of the Earth must be the centre of gravity for all things in the universe that are heavy. Furthermore, once a heavy object reaches the place towards which its natural movement sends it, it comes to rest. Applying this idea to the Earth, the inevitable conclusion is that the Earth must be at rest in the centre of the universe and that it cannot be moved except by some force strong enough to overcome its natural tendency. This argument was based on Aristotle's concept of 'natural' places and 'natural' movements. It is easier to see the validity of it if you realize that Aristotle thought of everything beyond the Moon being made up of something called aether, which was neither 'heavy nor light'.

4. The Sun-centred model did nothing to solve a problem astronomers had long been grappling with: the inequality of the seasons measured by the solstices and the equinoxes.

It would be inaccurate, and unfair to Aristarchus's contemporaries, to say that his Sun-centred model was suppressed because of their ignorance and closed-mindedness. The fact is, it was an inspired guess that we now have the means to know was right. But there actually was nothing coming from observation then to recommend it over what was the more orthodox view of the universe – the Earth-centred view that had been around for hundreds of years and that would be brought to its most sophisticated form in the work of the astronomer Claudius Ptolemy (not necessarily related to the Ptolemaic dynasty) four centuries after Aristarchus. Ptolemy's model would brilliantly explain the movements of the planets *if* the Earth is the centre – and, indeed, it solved the problems of astronomy *as they were perceived at the time* better than Aristarchus's model. Aristarchus's idea was a seed sown far too early, in a season in which it could not possibly germinate and take root. Ptolemy's Earth-centred astronomy would dominate thinking about the cosmos until the 16th century AD.

That is not to say that no further progress was made in ancient times towards understanding the heavens.

Hipparchus of Nicaea, who lived in the second century BC, was one of the most skilled astronomers the world has known, and he laid the foundation for much that was to follow. Hipparchus had at his disposal a priceless collection of Babylonian astronomical records – a legacy of Alexander's conquests – which he put to splendid use in his own astronomy, meticulously comparing the positions and patterns of stars and planets over the centuries with those he observed. Like Aristarchus and Eratosthenes, Hipparchus tried to find a way to calculate the distances and dimensions of the Sun and Moon. Part of his inheritance from the Babylonians was eclipse records spanning many hundreds of years. He also used a new line of reasoning, focused on the fact that there is no discernible change in the Sun's position against the background of stars when we move from one point to another on the surface of the Earth. To put that in more scientific language, there is no 'solar parallax motion'. Hipparchus worked on the assumption that observers on the Earth only just miss seeing solar parallax – in other words, that solar parallax is just below the threshold of visibility – and took it from there, with little success in terms of matching modern calculations.

One of Hipparchus's most impressive achievements – which came from comparing his own observations with observations made about 160 years earlier – was discovering the change in the relative positions of the equinoxes and the fixed stars. That is, if we look at the stars on the evening of the spring equinox, and then again on the evening of a spring equinox some years later, the stars will not be in the same position. In fact, they won't be in the same position again for 26,000 years! This phenomenon is the 'precession of the equinoxes'. Though Hipparchus couldn't discover its cause, he gave an accurate estimate of the rate of this change.

Of all Hipparchus's writings only one youthful, minor work survives. Next to nothing is known about his life or where he spent it, except that his name indicates that he must have hailed originally from Nicaea, in the northern part of what is now Turkey. Information about his accomplishments comes only from the references of others, mainly Ptolemy, but that evidence is sufficient to show that Hipparchus was an extremely fine astronomer and that he vastly improved observational techniques.

Where did Hipparchus stand in the competition between Aristarchus's model of the universe and the more orthodox one? Definitely pro-orthodox. Hipparchus was among those who did not accept Aristarchus's Sun-centred cosmos, and he influenced others to reject it. Hipparchus felt obliged to abide by the evidence of observation – observational astronomy was, after all, one of his fortes – and, as we have seen, observation didn't support Aristarchus and couldn't confirm the enormous distances required by the Sun-centred model. Hipparchus's own work contributed significantly to Ptolemy's later Earth-centred model of the cosmos. Some scholars even insist that Ptolemy's astronomy was by and large a re-editing of Hipparchus's, that Hipparchus was the genius and Ptolemy the textbook writer.

The Roman Pliny the Elder wrote of Hipparchus:

Hipparchus did a bold thing, that would be rash even for a god, namely to number the stars for his successors and to check off the constellations by name. For this he invented instruments by which to indicate their several positions and magnitudes so that it could easily be discovered not only whether stars perish and are born, but also whether any of them change their positions or are moved and also whether they increase or decrease in magnitude. He left the heavens as a legacy to all humankind, if anyone be found who could claim that inheritance.

'If anyone be found . . .'?

CHAPTER 2

Heavenly Revolutions
AD 100–1600

Now authorities agree that Earth holds firm her place at the centre of the Universe, and they regard the contrary as unthinkable, nay as absurd. Yet if we examine more closely it will be seen that this question is not so settled, and needs wider consideration.

Nicolaus Copernicus, *De revolutionibus orbium coelestium*

A few years ago, Harvard astrophysicist and science historian Owen Gingerich received a flyer in his mail offering a $1,000 prize for 'scientific proof-positive that the Earth moves'. A Mr Elmendorf, who posed this challenge, wrote, 'As an engineer, I am astounded that the question of the Earth's motion is apparently not "all settled" after all these years. I mean, if we don't know *that*, what do we know?'

Indeed, whether the Earth moves can hardly be classed as one of the great unsolved mysteries of science. Most of us have accepted since we were children that we live on a planet that revolves on its axis and orbits the Sun. We learned in school that Nicolaus Copernicus introduced this controversial idea in the 16th century and that some men were persecuted for believing

29

it. But in the end . . . 'all settled' . . . case closed. That was 400 to 500 years ago. What, we want to ask Mr Elmendorf, is the fuss about? And why has no one won the $1,000?

History and science turn out to be far more subtle and ambiguous than we were taught at primary school.

Certainly no scientific knowledge has a better claim to being 'Truth', with a capital T, than the knowledge that the Sun is the centre of our planetary system and that the Earth orbits it like the other planets. Yet our own contemporary science backs away and tells us that when it comes to proving whether there is an unmoving 'centre' and, if so, where it is, no one can make an air-tight case that any choice is *wrong*. Pick what you will, the Moon, Mars, the Sun, the Earth, your great-aunt's dining table – the options are infinite – and it's possible to come up with a successful mathematical description of our planetary system with that as the 'centre'. In fact we're being parochial if we limit the exercise to our planetary system. It would be possible to describe the entire universe using any chosen point as the unmoving centre – and no one could prove 'You're mistaken, that thing *moves*!'

The issue here, we must remind Mr Elmendorf and ourselves, is one of relative motion only. We can measure the motion of an object only in relation to other objects in the universe. We do best to picture everything in motion and nothing as being the centre. But we could, if we set our stubborn minds to it, choose the Earth as centre as our ancestors did and describe everything else correctly in relation to that centre, making a case that the Earth *is* the centre and the only thing that isn't moving. If our mathematics were good enough, it would be impossible for anyone to show our choice was wrong. Of course it would also be impossible for us to prove we were right, because we could transfer our allegiance to Venus, or the Sun, or Alpha Centauri, or a black hole at the heart of the Andromeda Galaxy, and make a case for any of those.

If we haven't given much thought to the implications of 20th-century science, we may be as chagrined as Mr Elmendorf to realize that because of the concept of relative motion, no one can provide knock-down proof that the Earth moves. *Did* we learn anything from Copernicus? Postpone that question for a moment, for relative motion isn't Mr Elmendorf's only problem. One tenet of science is that while an explanation can be extremely convincing and useful, none should ever be considered 'final' or 'proved', or 'Truth'. All scientific explanations are, in principle, open to revision and even complete rejection when better ones come along. Henri Poincaré, scientist and philosopher of science, was referring to this open-endedness of science when he wrote: 'If a phenomenon admits of a complete mechanical explanation, it will admit of an infinity of other [mechanical explanations] which account equally well for all the peculiarities disclosed by experiment.' Does this apply even to the motion of the solar system? Indeed, that was the example Poincaré used.

This deeper 20th-century insight (or, some may prefer, mere technicality) notwithstanding, every generation tends to believe devoutly in the finality of its own science, always for the same excellent reasons: what we experience presents us with puzzles. We put our trust in plausible solutions, of which we can say, 'Of course, that explains it!' We choose the explanations that seem to make the best sense of things as we know them, and of things as we believe future generations will probably know them (to the best of our ability to predict). After all, that is the most anyone can ask of science in any era. But it isn't final, unassailable truth. We criticize our ancestors for not bearing that in mind, while we continue to err in the selfsame manner.

Carrying our modern world-view along on a visit to the past is notoriously inadvisable, but, in this chapter and the next, doing so selectively would not be a bad idea. Leave behind scient*ism*, the popular belief that current science is final Truth. Instead, bring along a less naive scientific world-view that recognizes the open-endedness of science. Bring along the concept of relative motion. With those things in mind, you'll be prepared to appreciate and sympathize as two of the most brilliant intellects in history account for the observed movement of pinpoints of light in the heavens.

Almost no information whatsoever exists about the life or personality of Claudius Ptolemaeus, known to us as Ptolemy, except that he worked at Alexandria during the second century AD and died in about 180. There is no record of where he was born, and the name Ptolemy doesn't indicate he was a member of the ruling family. Like Eratosthenes, Ptolemy was interested in a wide range of subjects including acoustics, music theory, optics, descriptive geography and mapmaking. Some of his maps were still in use as late as the 16th century. More significantly, Ptolemy drew together, from previous ideas and knowledge and out of his own mathematical genius, an astronomy that would dominate Western thinking about the universe for 1,400 years.

Ptolemy inherited an intellectual tradition that placed an unmoving Earth at the centre of the universe and that insisted that all heavenly movement occurred in perfect circles and spheres. Among his contemporaries who thought about such matters, most had come to assume that the physical appearance of things must be taken into serious account when one tried to figure out the structure of the universe. To be believed, an explanation must 'save the appearances'. That may seem so obvious that it is hardly worth mentioning, but it isn't an assumption present in all cultures nor was it supported by all schools of thought in the Greek and Hellenistic worlds.

The 'appearance of things', for Ptolemy, included what Hipparchus and others had recorded in star catalogues. Ptolemy also brought to his task an in-depth knowledge of previous attempts to explain and predict planetary movement as it is seen from the Earth. The origins of the longing to understand that movement are lost in pre-history, but Plato, in the fourth century BC, focused it in the question: 'What are the uniform and ordered movements by the assumption of which the apparent movements of the planets can be accounted for?' The ancient attempts to answer him, beginning with his pupil, the mathematician Eudoxus of Cnidus, were ingenious. Science historians disagree about how much Ptolemy's work was a synthesis of some of these earlier ideas and how much it owed to his personal genius. Either way it's clear he was a brilliant mathematician, and his achievement is almost unparalleled in the history of science.

When trying to understand Ptolemy it helps to pay an imaginary visit to a fair or amusement park.

Look first at a carousel designed for very young children. The horses do nothing else but move in a large circle. We'll put a light on the head of one horse, switch off all other lights, situate ourselves at the centre of the carousel in such a way that we don't turn with the carousel, and set it going. The light circles us steadily, never varying in speed or brightness, never changing direction.

If the Earth were the centre and were not moving or rotating, and if all the planets were orbiting it in circular orbits, we could expect to see each planet as we see the light on this carousel. That is, roughly, the way we see the Moon and the Sun, though their movements include irregularities that foil any attempt to describe them quite that simply. But we definitely do not see the planets moving in this manner, and neither did ancient astronomers and stargazers. Even as early as Plato and Eudoxus, those who studied the heavens knew that a model with simple circular orbits centred on the Earth couldn't adequately explain what was going on up there.

To illustrate one particularly mysterious problem: on the darkened carousel, suppose we see the light move ahead for a while, pause, reverse, then move forward again. The pattern continues to repeat itself. How to account for this movement? Someone might suggest that the light isn't attached to a horse's head at all. Instead it's on the cap of the ticket-taker who is moving around among the horses. But the movement looks too regular for that. Not quite random enough. Try again. Suppose the light is at the end of a rope, and someone riding one of the horses is swinging the light around his or her head as one would a stone in a long slingshot prior to launching it. Putting that movement together with the overall circular movement of the carousel might explain the apparent reversing, as the light circles

Figure 2.1

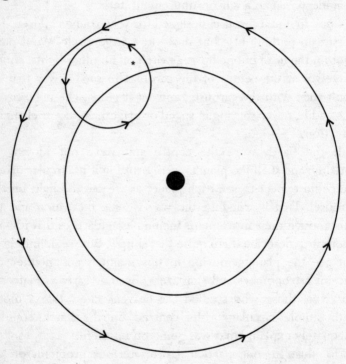

The carousel is rotating, and on its periphery, the smaller disc is rotating on its own axis so that the horse with the light on its head (asterisk) is chasing its tail. From the centre of the carousel it seems to us that the light moves forward, stops, reverses its motion a little while, stops again and moves forward once more.

toward the back of the rider's head. Or, if we want to stay with the notion that the light is on a horse's head, perhaps the horse isn't fixed directly to the floor of the carousel but instead is part of a mini-carousel attached near the edge of the large carousel. In other words, in addition to being carried around in the big circle of the carousel, the horse is also moving around in a smaller circle, chasing its own tail. This last scheme would be something like Figure 2.1. We could still accurately say that we are at the centre of the carousel and everything on it is moving around us. Also, all movement is in perfect circles, no matter how complicated and uncircular it may appear to our eyes.

Figure 2.2 is an idealized picture of the pattern such a light might trace in time-lapse photography taken from a helicopter hovering over the carousel. From our position at the centre (E) we wouldn't see the loops. The light would appear to go forward, then pause, then back up, then pause, then repeat the

Figure 2.2

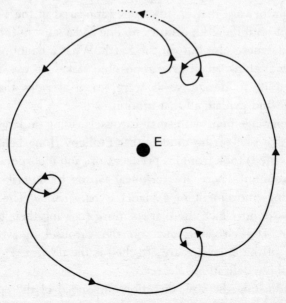

pattern. With the fair fully illuminated, it's easy to account for movement like that in terms of perfect circles, but it would require considerable mathematical insight to do so if we could see nothing but a few moving lights.

With the naked eye and a view of only a portion of the sky at any one time, those who studied the movements of the stars, Sun, Moon and planets before the invention of the telescope were left to account for what they saw in much the way we would have to do in our amusement park plunged into darkness. Tiny lights that move smoothly for a time, then pause and reverse, or seem to grow brighter and dimmer, to speed up and slow down. That's what Aristotle, Plato, Aristarchus and Ptolemy saw. That is also what Copernicus saw, for he too lived before there were telescopes. All they had beyond that were charts and catalogues of what their forebears had observed. Even assuming those notations were made with consummate skill and care (which was not always the case), given the lack of standardization and the discrepancy of calendars between one society and another, the interpretation of those records was not a straightforward matter. Telescopes appeared in the 17th century, but until the 20th century no one had a view of the planets from anywhere else but on the Earth. What a daunting task – one might think an utterly impossible task – it was to make sense of it, to design, as it were, carnival rides that could underlie and explain all that motion.

The arrangement with small carousels riding on large carousels is like one of the models that Ptolemy (long before such rides existed) took from his predecessors and incorporated into his astronomy. There are technical words to describe it: the 'reversing' movement of a planet is *retrogression*. The smaller circles around which the planets move (chasing their tails) are *epicycles*. The circumference on the carousel at which the centres of the epicycles are attached is the *deferent*. Figure 2.3 makes these definitions clearer.

By adjusting the size, direction and speed of the epicycles,

36

Figure 2.3

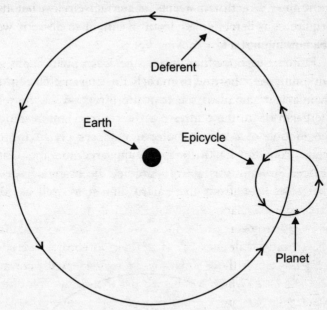

The planet moves in a small circle (epicycle) while that small circle moves along a large circle (deferent). The deferent is centred on the Earth.

it's possible to account for many irregularities in the way the planets move as viewed from the Earth in an Earth-centred system. It's also possible to explain irregularities in the movements of the Sun and Moon and variations in a planet's brightness. The epicycles place the planet sometimes closer to us, sometimes further away.

A second device that Ptolemy inherited from earlier astronomy and liked better as an explanation for the Sun's movement had the planet or the Sun orbiting in a circle with the orbit centred not precisely on the Earth but on a point a small distance away from the Earth, as in Figure 2.4. The technical name for the displaced circle is the *eccentric*. In this model also the planet will be closer to Earth in one part of its orbit than in another, which makes sense of variations in its brightness when viewed from Earth, and also of apparent changes in its speed.

Ptolemy used another device, the *equant*, which was original with him, not inherited from earlier astronomy. The equant was an imaginary point in the opposite direction (as viewed from the Earth) from the centre of the eccentric. This invention kept the motion of a planet uniform with respect to the equant, while from the Earth its speed appeared to vary. Using this device, Ptolemy was able to predict the changing speed of a planet (as seen from the Earth) almost as well as Johannes

Figure 2.4

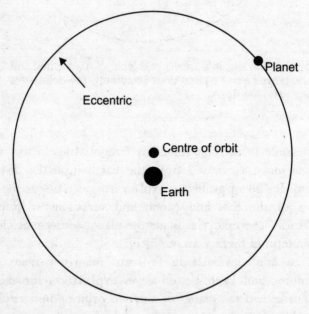

In this model, the circular orbit (the eccentric) is centred on a point a small distance away from the Earth.

Kepler would do in the 17th century. Kepler's is the way modern astronomy solves the problem, but Ptolemy's solution came close to being its geometric equivalent.

Ptolemy combined these devices – epicycle/deferent, eccentric, equant – in an elaborate and highly successful model of heavenly movement. To a degree of accuracy that astounds us who believe we know a much better way of thinking about it, his astronomy predicted and accounted for the observed movements of the five planets known in his time and of the Sun and the Moon, without removing the Earth from its position as unmoving centre. Ptolemy centred the seven orbits not on one point but on seven different points near the Earth. Adjusting epicycles, deferents, eccentrics and equants, he succeeded in doing what his forebears had hoped for for so long – devising a system that could explain all the observed movement in the heavens in terms of spheres and circles.

When ancient and medieval astronomers, and also early Copernicans, spoke of 'spheres', they weren't referring to the planets themselves. Since Aristotle first introduced the idea, long before Ptolemy, most astronomers had visualized the movement of each planet as taking place within the confines of its own invisible 'crystalline sphere'. Not everyone agreed as to the nature and mechanics of these spheres.

In order to understand this concept, you must not imagine a flat drawing of circles within circles, but picture instead a three-dimensional glass object made up of transparent (invisible) spheres-within-spheres-within-spheres, nested like Russian dolls – or like the bubbles within bubbles that skilled bubble-pipe blowers can produce. In the Ptolemaic arrangement, the Earth, at the centre, is surrounded by a sphere within which the Moon moves, and both of these are surrounded by a larger sphere representing the level at which Mercury moves, which in turn is surrounded by a still larger sphere representing the level at which Venus moves. Surrounding these three spheres is a sphere representing the level at which the Sun moves, and so on and so forth – larger and larger nested spheres representing the levels at which each of the other planets moves. The outermost sphere represents the level of the stars.

In Ptolemaic astronomy, each sphere leaves enough room for all that planet's epicycles. However, it leaves room for no *more* than that. Ptolemy took from Aristotle the notion that the cosmos is a plenum. In other words, it is 'full' in the sense that there can be no empty spaces between the spheres of the planets. Unlike Russian dolls or bubbles from a bubble pipe, the arrangement leaves no area that is not part of a sphere. The borders of the spheres 'touch'. For instance, the sphere of Jupiter, defined by Jupiter's nearest and furthest distances from the Earth, has an 'inner border' that is right up against the 'outer border' of the sphere of Mars – and so on and so forth with each planet's sphere and the spheres of its neighbours. Ptolemaic astronomy saw the spheres as pushing one another along. Movement originating in the sphere of the stars was thus transferred to the spheres of the planets, causing their circular movement.

In his most famous work, the *Almagest*, Ptolemy was chiefly concerned with predicting the positions of the heavenly bodies, not their distances. However, working from Ptolemy's planetary scheme in that book, it's possible to calculate the ratio between the planets' greatest and least relative distances from Earth, and Ptolemy was later to insist that this could be transformed into *absolute* distances. Ptolemy did the calculation first for the Moon and found that the least distance to the Moon was 33 times the radius of the Earth (that radius was known from Eratosthenes's measurement), and the greatest distance 64 times the Earth's radius. In a work that came after the *Almagest*, called *Planetary Hypotheses*, he applied the same thinking to the planets and to the Sun, taking the scheme that had been largely of mathematical significance in the *Almagest* and treating it as 'real' in terms of absolute distances. Ptolemy was unable to make it all work out without any glitches, but a strong argument in favour of his method nevertheless was that, with only small discrepancies, the results were largely in agreement with results of the study of eclipses.

As Ptolemy summed up his findings: the distance to the borders of the region 'of air and fire' (the 'sublunar' region) is 33 times the radius of the Earth's surface ('the spherical surface of earth and water'). At that radius, the sphere of the Moon begins. Its further border is 64 times that of the radius of the Earth (this was roughly where Aristarchus placed the Moon), and that is where the sphere of Mercury begins. And so on from there. The outer border of the sphere of Mercury is 166 times the Earth's radius; Venus, 1,079 times; the Sun, 1,260 (which agreed with Aristarchus's erroneous measurement to the Sun); Mars, 8,820; Jupiter, 14,189; Saturn, 19,865. That radius is where the sphere of the fixed stars begins. When it came to the absolute distance to the fixed stars, Ptolemy wrote: 'The boundary that separates the sphere of Saturn from the sphere of the fixed stars lies at a distance of five myriad myriad and 6,946 myriad stades and a third of a myriad stades.' That comes out to 569,463,333 stades. Translated into modern miles, roughly 50 million miles. Today we measure the distance to the nearest star as over 25 trillion miles.

The term 'Ptolemaic astronomy' is a bit of a misnomer. It doesn't refer to a specific set of solutions coming from Ptolemy himself or any one of his successors. Instead, it means the combination and recombination of Ptolemy's devices as astronomers used them over the centuries. It is ironic to note that modern physicists and astronomers – who believe so firmly in a moving Earth – can recognize the power of these devices better today than 16th-century Ptolemaic astronomers could. It takes understanding acquired through Copernican astronomy, as well as more highly developed mathematics than were available to ancient and medieval scholars, to appreciate how successful the older system potentially *was*.

Ptolemy's work wasn't lost with the decline of ancient

civilizations. It had reached Baghdad in the eighth century AD, and there it was translated into Arabic. The *Almagest* is an Arabic title meaning 'The Greatest'. From the eighth to the 13th and 14th centuries, the development of 'Western' mathematics, astronomy and astrology (astronomy and astrology were one and the same discipline then and continued to be even as late as the 17th century) took place not in Latin Europe but in the Middle East, north Africa and Moorish Spain. Most scholars there who were responsible for this growth were Islamic, but not all, for these societies were cosmopolitan and tended to be tolerant of non-Islamic thinkers. Mathematicians and astronomers moved beyond the mathematical methods of the Greeks and Ptolemy, constructing observatories in Baghdad, Cairo, Damascus and other leading cities. These observatories contained no telescopes, of course, but they did have intricate apparatuses – some inherited from the ancients and others that were newer inventions – to help plot the movements of the planets. Astronomical instruments came both in large bulky sizes, made of masonry, and in smaller portable models.

Islamic astronomers seem never to have questioned the Earth-centred model. However, they did criticize Ptolemy on details, particularly for the use of the equant, which, of his devices, seemed nearest to violating the requirement of uniform circular motion. When Ptolemaic astronomy later became a well-established field of study in European centres of learning, European scholars did not forget their debt to Islamic scholarship. Copernicus mentioned some of his Islamic scientific forebears by name in his book *De revolutionibus*.

There were attempts among Islamic astronomers to estimate the distances to the planets and the stars. One man who tried was Al Fargani, a ninth-century Arab astronomer who attempted to gauge the sizes of the spheres and worked out relationships among these sizes. As Ptolemy had done, he allowed each sphere to be large enough to contain its planet's epicycles. Starting from a measurement of the Earth's radius of

3,250 Roman miles, he then used these relationships to calculate the distances to all the known planets and to the sphere of the stars. His measurement of the distance to the stars had them more than 75 million miles from the Earth, half again as far as Ptolemy put them. The modern distance measurement to the nearest star is in the neighbourhood of a million times greater than Al Fargani estimated to the sphere of the stars, but 75 million miles from the Earth was still a huge distance, and the sublunar (lower than the Moon) region in Al Fargani's universe was tiny indeed by comparison.

Ptolemaic astronomy, preserved and improved by Islamic scholars, filtered slowly into Latin Europe beginning as early as the 11th century, but European scholars were much quicker to assimilate other ancient thinking and much more strongly influenced by it. The translation of Aristotle into Latin in the 12th century had a profound impact. Scholars came to revere him not just as *a* philosopher, but as '*the* Philosopher' – no other identification required – and the final authority on science and cosmology. With the passage of time, Aristotelian cosmology as interpreted by medieval scholars merged with medieval Christian thought and, somewhat later, with Ptolemaic astronomy – which agreed with Aristotle in many, but not all, respects. Scholars, who at that time were also clergy, after a period of disagreement and debate settled on ways to reconcile the Bible with Aristotle. In order to accomplish this, they gave scripture a less literal, more metaphorical reading. They also eventually succeeded, to their satisfaction, in resolving contradictions among Ptolemy, Aristotle and other ancient scholars to produce a coherent body of philosophical, religious and scientific thought.

Aristotelian/Ptolemaic astronomy provided a visual, geometric structure for abstract medieval Judaeo-Christian concepts, with all the rest of creation centring on that stationary Earth that is the home of humankind – a singularly unsavoury place in Aristotle's philosophy, fallen from God's grace in

Judaeo-Christian teaching, and minuscule according to Al Fargani. By the 13th century, educated Europeans were taking it for granted that this world-view represented reality, and Dante, in the 14th century, both reflected and reinforced it in his *Divine Comedy*. He described a descent through the nine circles of hell towards the centre of the Earth — the vilest point in the universe — and an ascent through the celestial spheres in which the planets move, to the throne of God.

By the time Copernicus was born in the 15th century, the simple Aristotelian picture of the cosmos and complex Ptolemaic astronomy (as received from the Arabs) had been mingling for at least two hundred years in the minds of European thinkers. Decked out in poetry and metaphor by Dante, this Aristotelian/Ptolemaic/Judaeo-Christian view of the universe now not only described and predicted heavenly movement with reasonable accuracy and served as a map of the physical universe, but had also come to portray the human condition and the geography of the spiritual universe: among all creatures, only humans, fallen though they were, combined the material and the spiritual. Humanity was torn between the two, stuck on the debased squalid central Earth but always within view of and reaching out for the holy, pure and changeless realms beyond.

At the same time, the notion that the Aristotelian/Ptolemaic picture of the universe might be wrong was not completely absent from European thought. In the 14th century it took the form of a challenge to Aristotle, not to Ptolemy, when Parisian Nicole Oresme wrote a commentary criticizing Aristotle's conclusion that the Earth didn't move. Oresme didn't propose that the Earth moved in orbit. There is no suggestion that he thought of that. Actually, he didn't even conclude that it rotated, only that Aristotle had not proved it *didn't*. A century later, Cardinal Nicholas of Cusa, a scholar who died nine years before Copernicus's birth, also suggested that the Earth was not lying motionless in the centre of the universe. Nicholas of Cusa, however, failed to suggest another centre. The next and far stronger challenge to Aristotle, Ptolemy and Dante was to come from Copernicus.

Today the Polish city of Torun on the banks of the Vistula River has large modern buildings, but there also still exists a section of Old Town whose narrow cobbled streets can't have changed much since 1473, when Mikolaj Kopernik was born here. It was common practice in scholarly circles to Latinize one's name. Mikolaj Kopernik became Nicolaus Copernicus.

Poland has had a tumultuous history and has often been divided and dominated by foreigners, but the period during which Copernicus lived was a relatively peaceful, prosperous time. Copernicus's father and grandfather evidently were merchants – well off if not spectacularly wealthy – and his mother came from a prominent Torun family. When Copernicus was 10, his father died and his mother's brother took charge of Copernicus's upbringing and education. That uncle rose to become Bishop of Warmia, an influential position from which to advance the careers of his nephews.

When Copernicus was 18 and his brother Andrzej (Latinized to Andreas) 20, the two young men enrolled in the Jagiellonian University in Krakow, one of the most celebrated seats of learning in Europe and renowned for its astronomy. One of Copernicus's purchases while he was a student there was a set of the Alphonsine Tables. These tables, used for finding the positions of the Sun, Moon and planets based on Ptolemy's theories and Islamic observations, had been computed in the 13th century at the behest of King Alphonso of Castille. Copernicus's copy still exists. Judging from the stains, he used it a great deal.

When Copernicus was 22, he and Andreas 'walked across the Alps' (tradition tells us) to Italy to continue their studies. The influential uncle had taken his Doctorate of Laws at Bologna, Italy's oldest university and particularly famous for its faculty of law. He hoped to draw Copernicus's interest away from astronomy and into law. However, Copernicus didn't pass up the opportunity to get to know the leading scholars of astronomy and astrology at Bologna.

One eminent teacher there, Maria de Novara, not an astronomer but a mathematician, may have influenced him profoundly. Novara was a Neoplatonist and Copernicus had already expressed some sympathy for Neoplatonic opinions. Neoplatonism was a framework of thought that stressed the need to discover simple mathematical and geometric reality underlying all the apparent complexity of nature. Neoplatonists, following in the tradition of Pythagoras, also saw the Sun as the source of all vital principles and energies in the universe. Novara himself insisted that nothing so complex and cumbersome as Ptolemaic theory could correctly represent the truth.

Copernicus was mathematician enough to find the complications of that astronomy unwieldy and annoying. But somewhere along the line – and we can't know precisely when – Copernicus went beyond that annoyance to become thoroughly convinced that a simpler model of the universe was much more likely to be a correct model. The answer to the question, why *now* a challenge to Ptolemy, before there was any compelling observational reason for it, may lie partly in Copernicus's association with Novara.

Due to their uncle the Bishop's influence, Copernicus and his brother had received appointments as canons of Warmia, positions involving both Church and civil authority. Nevertheless, the Chapter there was pleased to extend Copernicus's leave of absence for educational purposes when he announced his intention to study medicine. Doctors were in short supply. So Copernicus set off again in 1501, this time for the University of Padua with its famed faculty of medicine. There he did indeed expend much effort studying medicine, though, mysteriously, the degree he eventually received was a Doctorate of Canon Law from the University of Ferrara. The explanation for the change of venue may be that he didn't know anyone at Ferrara and hence could avoid the expense of the expected celebration party.

Copernicus returned to Poland in 1503 and – surprisingly, after such a cosmopolitan start – never left again. He began to build a distinguished reputation as a doctor, reportedly saving many lives later during a severe epidemic in 1519. Reading his notes, some of which arguably represent the best of medical practice at the time, leads one to wonder how this happened.

From 1503 until 1512, Copernicus, now in his thirties, lived at Lidzbark Castle, the seat of his uncle the Bishop of Warmia. He served as his uncle's secretary and personal physician and became a force in the politics of Warmia, putting to use the legal education he had received at Bologna. He also found time for astronomical observations and kept painstaking records of them. About 1507, while still a member of the court at Lidzbark, he wrote a book in which he claimed that the Ptolemaic model was wrong. The book was short, about 20 handwritten pages. In that form (it was not printed) it circulated at first anonymously and with an unrecorded title among Copernicus's scientific acquaintances. The later title was *Nicolai Copernici de Hypothesibus Motuum Caelestium a se Constituis Commentariolus*, usually shortened to *Commentariolus* (*Commentary*).

Sun-centred astronomy as Copernicus first proposed it in *Commentariolus* didn't by any means solve all the problems still nagging Ptolemaic astronomy. Trying to eliminate the equant, which he found particularly offensive, and keeping the orbits circular, he was forced to use epicycles to account for the movement of the planets. There is little mathematical reasoning in *Commentariolus*. It is hardly more than a sketch. Nevertheless, the leap that Copernicus made was extraordinary. He claimed that by putting the Sun in the centre it would be possible to explain the heavens more simply and logically than Ptolemaic astronomy had done.

In *Commentariolus* Copernicus still visualized the universe in terms of spheres. The Sun rather than the Earth is at the centre of his arrangement, and there is a sphere representing the level (from the Sun) at which the Earth moves, encasing the Sun and the spheres of Mercury and Venus. The small sphere in which the Moon moves is the only sphere that has the Earth as its centre.

Copernicus proposed seven 'assumptions':

1. All celestial spheres do not have only one common centre.
2. The centre of the Earth is not the centre of the universe, but the centre of the Earth *is* the centre of gravity and of the Moon's sphere.
3. All spheres (except that of the Moon) revolve around the Sun, as though the centre were the Sun, so the centre of the universe is near the Sun.
4. The firmament of stars is extremely far away. The distance from the Earth to the Sun is insignificant when compared with the distance from the Earth to the firmament.
5. What we see as motions in the firmament of stars are not its motions, but those of the Earth. The Earth with its adjacent elements, air and water, rotates daily around its poles while the firmament remains motionless.
6. What we see as motions of the Sun are not its motions, but the motion of the Earth and of the sphere with which we revolve around the Sun in the same manner the other planets revolve in their spheres.
7. What appears to us as retrograde and forward movement of the planets (the backward and forward motion) is not their motion, but that of the Earth. The Earth's motion alone is sufficient to explain many different phenomena in the heavens.

Copernicus went on to write: 'The highest is the sphere of the fixed stars, containing and fixing location for everything. Below it is Saturn, followed by Jupiter, then Mars; below it the sphere in which we move, then Venus and finally Mercury. The lunar sphere revolves around the centre of the Earth.'

It would seem that Copernicus had delivered a bombshell, but there was no explosion. Few people outside Poland heard about his book. His suggestions went largely undiscussed and unchallenged. The Catholic Church became aware of him but evidently didn't regard his ideas as a threat. The Church had for at least two centuries been adopting a tolerant, hands-off attitude towards ideas that challenged traditional astronomy – even suggestions such as that from Nicholas of Cusa, one of its own cardinals. If any dyed-in-the-wool Ptolemaic astronomers recognized Copernicus as a potential challenge, they may have sagely decided that the best defence against the new model would be not to respond to it at all. Let it die in oblivion.

In 1512, Copernicus's life story took a new turn with the death of his uncle. He left the castle at Lidzbark and went to live at Frombork, for he was still a canon of Warmia and Frombork was the seat of the Chapter. There, in a tower adjoining the cathedral, he would do his most important work. He described his residence there as 'the remotest corner of the Earth', which seemed not to bother him unduly or deflect him from his scholarship. Copernicus was a modest, quiet man who saw little objection to having himself, and the Earth, removed from the centre of things.

During his years at Frombork, Copernicus made a number of astronomical observations – less accurately than the Greeks had done, for Copernicus was not particularly skilled at observational astronomy. He also began or continued work on a second book. However, although the death of his uncle seemed likely to leave Copernicus with more time to devote to his astronomy, there were serious distractions. He became involved in negotiations with the Teutonic Knights, one of the forces that periodically threatened to tear Poland apart. When diplomacy failed and fighting broke out, Copernicus moved to Olsztyn, though he was too stiff-necked to retreat to Gdansk with the rest of the canons and remained at Olsztyn, now under siege, to organize resistance. When the siege was finally lifted and the fighting stopped, he headed up relief operations. Copernicus became involved with problems of economics during the recovery after this conflict, and he wrote a short, insightful tract proposing currency reforms. It's surprising to learn that had Copernicus not introduced Sun-centred astronomy, he might still be remembered as a minor historical figure in the field of economics.

Copernicus eventually got back to Frombork, where his duties as an administrator and doctor must still have consumed the greater part of his time. Nevertheless, he resumed work on his book. By now that may have been substantially completed, but he was a meticulous man, worrying over details, trying to confront and solve every problem presented in the astronomical data he had. He was frustrated for example by the discovery that the Earth's whole orbit seemed to oscillate. He called these oscillations 'trepidations' and tried to account for them. Later astronomers found that these trepidations were an illusion. Copernicus had been worrying himself over nothing but the result of bad data. One of our own century's great astronomers, Fred Hoyle, writes of his predecessors in his book *Nicolaus Copernicus: An Essay on His Life and Work*:

> The early astronomers could not know which problems they could hope to solve. Perforce they had to take a shot at everything. And because insoluble problems were mixed up with soluble ones their task in dealing with the soluble ones was made all the harder. This must always be remembered in attempting to understand the difficulties which beset Ptolemy and Copernicus. Both expended much effort in attempting to understand the Moon, whereas they would probably have gone further with less effort if they had ignored this problem.

Copernicus's friends urged him to publish his book while he was still fretting about its not being quite ready. We are reminded of Johannes Brahms carrying the completed manu-script of his first symphony around with him in his coat pocket, unable to bring himself to relinquish it to the public and wanting to be dead certain none of his friends did it for him. Copernicus was about to insist on a vision that disagreed with what nearly every intelligent, educated person had been think-ing for millennia. He had become a well-known and well-respected astronomer and had no desire to appear an eccentric lunatic. His conflict with the astronomers of the past was a battle we think of as being fought over the big picture, but for Copernicus it was a battle fought in terms of technical and mathematical minutiae. Until he could work out these details to his own satisfaction, he could not feel he had succeeded. Even with all this concern, there *were* loose ends. Copernicus was convinced that his rearrangement of the universe could yield a far more harmonious and effective astronomy, but he failed to demonstrate this improvement as effectively as he hoped to do.

Copernicus had other worries dogging him besides his book. He was in his sixties now, no longer in the prime of health. Gnapheus, a minor playwright, produced a comedy, *The Wise Fool*, that mocked Copernicus. There were rumours of some ill-considered, scoffing dinner table remarks about moving the Earth coming from Martin Luther. Closer to home, Copernicus became embroiled in a demeaning squabble with the new Bishop of Warmia. Reports differ as to whether Copernicus had opposed or supported this man's election, and whether it was a vindictive move when the Bishop undertook to remove Coper-nicus's housekeeper Anna Szylling, a widow and distant relation on Copernicus's mother's side. She was a handsome, cultured woman whose presence in Copernicus's house the Bishop (reportedly no paragon of morality himself) deemed suspect and unsuitable. Copernicus resisted for over a year but finally agreed to her departure.

Despite these griefs and distractions, in the late 1530s, 1,700 years after Aristarchus, Copernicus was at last drawing to a finish the book that ultimately would lead to the vindication of that ancient astronomer and be one of the most significant watersheds in human intellectual history – *De revolutionibus orbium coelestium* (*Concerning the revolutions of the heavenly orbs*). Copernicus wrote out the manuscript himself in longhand, as he had done with *Commentariolus*. This time there were more than two hundred pages. The book would elaborate on the sketch he had given in *Commentariolus*, his primary assertion again being that the Sun, not the Earth, must be considered the centre of the system.

While Copernicus was still labouring on *De revolutionibus*, rumour got out that something radical was in the making at Frombork. The circulation of *Commentariolus* was small, but it was significant, and Copernicus's friends, particularly Bishop Tiedemann Giese, were spreading the word enthusiastically. Already in 1533, Pope Clement VII requested that his secretary explain these new Sun-centred theories to him. In 1536, Nicolaus Schönberg, Cardinal of Capua, wrote to Copernicus asking about his theories, and Copernicus sent him some explanations and tables. Cardinal Schönberg moved firmly into the Copernican camp. He urged Copernicus to allow his book to see the light of day and offered to pay for its publication and printing. Unfortunately, the Cardinal died before he could make good on his offer, but Copernicus mentioned this strong encouragement as well as that of Bishop Giese in the dedication of the book. However, it was not a Catholic churchman but a young mathematician named Rhaeticus, from Protestant Wittenberg, who at last persuaded Copernicus to publish his work.

Rhaeticus hadn't had an enviable adolescence. When he was a teenager, his father was beheaded as a sorcerer, and Rhaeticus, who had previously been Georg Joachin von Lauchen, changed his name. Rhaetia was the province where he was born. Now he was a junior professor at the University of Wittenberg, and he was deeply impressed with what he heard about Copernicus's ideas. In 1539 he travelled to Frombork to meet Copernicus in person. Rhaeticus evidently didn't lack for courage, because Wittenberg, his university, was the centre of Lutheranism, while Warmia, where Copernicus was a canon of the cathedral, was Catholic and profoundly anti-Lutheran. But the two men, one 66, the other 22, seem to have hit it off splendidly, and Rhaeticus's visit stretched on for two years. He paved the way for *De revolutionibus* by publishing a short volume of his own summarizing Copernican theory. His book was favourably received. Finally Copernicus agreed to publish.

The tale, as it continues, is a convoluted one. The favoured version is that Rhaeticus had to return to Wittenberg before Copernicus was quite prepared to part with the manuscript of his book. Rhaeticus took with him only some early mathematical chapters. A little later, with Copernicus's consent, Bishop Giese sent the completed manuscript on to Rhaeticus. Rhaeticus took it to a Nuremberg publisher, intending to keep close watch over its printing. However, he was then appointed to a new position with a higher salary in Leipzig and delegated what remained of the proofreading to Andreas Osiander, a Lutheran clergyman who was more nervous about possible religious reactions than Rhaeticus was. The Catholic Church had had nothing to say one way or another, except for the support of Cardinal Schönberg and Bishop Giese, but there had been those adverse remarks from Luther. Osiander urged Copernicus to protect himself by writing a preface saying that his theory was intended to be taken hypothetically, not as a truth claim. Copernicus refused. He dedicated his book to Pope Paul III, a scholar interested in science. Osiander decided to write, himself, the preface Copernicus wouldn't write. He left it unsigned, probably because he feared that his own anti-papal reputation would cast suspicions on Copernicus.

On 24 May 1543, about a month after the printing of *De revolutionibus* was completed, Copernicus died. Tradition has it that he saw the printed book. He had had a stroke and was bedridden, perhaps unconscious, so there is some doubt whether that story is true. Was he aware enough to learn that Osiander had written, anonymously, the preface he himself had refused to write, saying his new scheme was only hypothetical and containing the warning: 'Beware if you expect truth from astronomy lest you leave this field a greater fool than when you entered'? If Copernicus knew of this, there is no record of his reaction.

Even after Copernicus's prodigious effort and foot-dragging about publication, *De revolutionibus* did not make the case for Sun-centred astronomy as effectively as he had hoped. There were many loose ends. Though he had been able to eliminate the use of the equant, he'd still had to use epicycles and eccentrics to explain the movements of the planets. The result was hardly less complicated and cumbersome than Ptolemaic astronomy.

However, the new system definitely had some things going for it. The new arrangement of the heavens allowed Copernicus to come at the problem of the mysterious 'reversing' or 'retrograde' movement of the planets in an entirely fresh way. Retrograde movement occurs when a planet is in 'opposition', meaning that it is on the opposite side of the Earth from the Sun. (Only Mars, Jupiter, Saturn and the other outer planets discovered since Copernicus's time can be in opposition. Venus and Mercury, whose orbits are closer to the Sun than Earth's, can never be in opposition.) Most of the time, the planets move from west to east against the background of stars. However, around opposition a planet appears for a while to move east to west. Ptolemy had used epicycles to solve this problem. In the Copernican model, with all planets including the Earth orbiting the Sun, when one of the planets is in opposition, the Earth catches up and runs ahead of the other planet. For an analogy, imagine two racing cars, one on an inner track and the other on an outer track. We are riding in the one on the inner track. The stadium is completely dark except for a light on top of the car on the outer track and some distant streetlights way beyond that. When our car catches up with that car and moves on ahead, the light will appear to us (against the background of streetlights) to backtrack. If the motion of our car is so constant and smooth that we believe we are standing still, we'll conclude that the other car has stopped for a moment, reversed, stopped again, and continued its forward motion. See Figure 2.5.

De revolutionibus also made sense of the fact that Mercury and Venus never stray far from the Sun. Ptolemaic astronomy had used deferents and epicycles to unravel this mystery. In Copernicus's model, with the orbits of Mercury and Venus lying within the Earth's orbit (they are both closer to the Sun than the Earth is), there is no mystery. Observers on Earth couldn't possibly see these planets anywhere else but near the Sun. This explanation of the orbits of Mercury and Venus was one success of Copernicus's model that many of his contemporaries could immediately appreciate.

Copernicus also addressed the ancient objections to the idea that the Earth rotates on its axis and moves in orbit. Judging from modern knowledge about the availability of certain books during his lifetime, most scholars conclude that when he wrote *Commentariolus* he probably didn't know about Aristarchus's suggestion that the Sun rather than the Earth is at the centre of the universe. However, it's clear from statements in *De revolutionibus* that by the time he wrote that book he had heard about it. Copernicus explained that the Earth carries its atmosphere with it as it spins, and insisted, as Aristarchus had done, that the fact that we observe no stellar parallax proves the stars are extremely far away. Al Fargani had estimated the distance to the sphere of stars to be more than 75 million miles. Copernican astronomy required that the distance be 75 times as great as that. The vast amount of empty space this leaves between the sphere of Saturn and the sphere of the stars was, however, not reflected in an illustration from *De revolutionibus* (see illustration

Figure 2.5 Copernicus's explanation for the retrogression of a planet

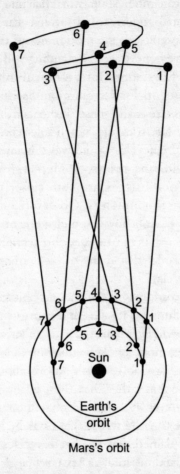

The Sun is in the centre. The inner ring is the Earth's orbit. The outer ring is the orbit of the planet. Place yourself at 1 on the Earth's orbit, then at 2, and so forth, as the Earth moves in its orbit. The line drawn through the corresponding number on the planet's orbit shows where the planet is in your line of sight in each instance. The squiggle at the top of the drawing shows the pattern these changes of position (of both Earth and planet) will produce against the background of distant stars, and why the planet will seem to 'reverse'.

section), showing Sol, the Sun, at the centre of the universe, with seven planets in their spheres around it. Beyond those spheres is another named 'Stellarum Fixarum Sphaera Immobilus', the 'immobile sphere of the fixed stars'. Copernicus, unlike Ptolemy and Ptolemaic astronomers, thought that the stars were stationary.

In 1551, eight years after Copernicus's death and the publication of his book, the first handy-to-use tables based on Copernican theory appeared. They were a substantial improvement on the Ptolemaic tables (partly because there had been no new tables for a very long time), but they were far from completely accurate, because astronomers were still so dependent on Ptolemaic observations.

No period in the evolution of thought about the universe and humankind's place in it has been more complicated or more ultimately decisive than the century and a half following the publication of *De revolutionibus* in 1543. Historians speak of a scientific 'revolution'. A hundred and fifty years is a long, slow-moving revolution by political standards, but no political revolution has been more profound. It is interesting that the shot eventually heard round the world took so long to reverberate and came from a narrowly technical book that only highly trained astronomers and mathematicians could understand, and whose author didn't make his case very effectively. Yet over the course of a century and a half, in the minds of an increasing number of people, an Earth-centred universe became a Sun-centred infinite universe, and science became, for many, the chief arbiter of truth.

One reason for the initial dearth of reaction was that even the literate, educated public lacked the expert knowledge to understand *De revolutionibus*. Among those who could understand it, few thought Copernicus himself had meant his arguments as a truth claim. Osiander's preface, which many assumed Copernicus had written, had something to do with this impression, but also, in the Ptolemaic tradition, new theories of a mathematical astronomer were normally intended as useful models for making predictions about planetary positions, not proposals for changing humankind's view of reality. The happy result for Copernican astronomy was that most scholars had found some of Copernicus's mathematical techniques too useful to discard by the time any actual opposition to his central thesis emerged. On the strength of these mathematical techniques Copernican theory infiltrated the scholarly world, but the subversion of Ptolemaic astronomy was a slow process, and historian John Hedley Brooke points out that we can identify only 10 people in the years between 1543 and 1600 as 'pro-Copernican' to the extent of stating that they actually believed that the Earth moved.

Some scholars chose to accept Copernicus's proposal that the Earth rotated on its axis but not his proposal that it was in orbit around the Sun. Others were willing to have the planets orbiting the Sun, while the Sun itself revolved around the Earth, carrying the planets along. This scheme came from the great Danish astronomer Tycho Brahe and also from Nicolai Reymers Baer (better known as Ursus, Latin for bear), the official mathematician of Holy Roman Emperor Rudolf II. There were accusations and counter-accusations of plagiarism between the two men. In view of the concept of relative motion, this model was not ridiculous at all. In fact it is the geometric equivalent of the Copernican model.

As word of the book and rumours of its contents spread, not all the debate about *De revolutionibus* took place on strictly astronomical or mathematical grounds or among people who understood it. Reactions from outside astronomy were mostly negative, some of them vehemently so. Copernicus's Sun-centred astronomy challenged a cosmic structure that men and women had taken for granted since centuries before Christ. Deep reverence for Aristotle still pervaded the intellectual world. Copernicus had insisted the Earth was one of the planets, but didn't we know from Aristotle that the Earth was made of different stuff from the planets? The planets were made of a fifth element that was not to be found in the corrupt realm within the Moon's orbit.

Tycho Brahe gave the world some fresh cause to doubt the Aristotelian picture of the universe by demonstrating that a nova in 1572, the great comet of 1577, and five more comets during the next 20 years were all further away than the Moon. This was a shocker, because Aristotle had taught that birth, death and change took place only in the sublunar part of the universe, and that beyond the Moon were only the eternal and unchanging celestial spheres. Tycho's discovery might seem a strong vote for Copernicus, but it fitted equally well with Tycho's concept of all the planets orbiting the Sun while the Sun orbits the Earth.

Though the problem of having humankind dethroned from the centre of the universe would frequently crop up as an objection to Copernicus's system, it was also possible to see this move as an innocuous or even a fortuitous one: Copernicus had argued that the fact that observers on Earth detect no stellar parallax as Earth travels in its orbit must mean that the stars are extremely far away. So in the Copernican system humans are still very close to the centre. To all intents and purposes, with the distances being so great, we are still *at* the centre. Others called attention to the fact that the centre of everything in the Aristotelian scheme was not such prime real estate after all. Beyond the Moon's orbit was perfection; beneath it, corruption, at the centre of which was the Earth. It was the realm of degradation and change. Christianity had come to associate it with fallen humanity. Satan lived in hell at the core of the Earth. Why cling so tenaciously to the notion that we live in the armpit of the universe? How welcome to find ourselves well out of that, moving around, breathing purer air! There was even, reportedly, criticism of the Copernican system for elevating man *above* his true station.

The question of possible extra-terrestrial life arose and became a topic in religious discussions and writing. If Earth is a planet, might not other planets have inhabitants too? Had they fallen from God's grace as we had? Did Christ die for them as well? Though this problem was a matter for conjecture and concern, religious scholars as a whole didn't decide it was a reason for rejecting the Copernican system, though some very vocal opponents mouthed it about a great deal. Another issue: before Copernicus, heaven and hell were considered to be not at all in the same neck of the woods. Heaven was beyond the outermost sphere. Hell was deep within the Earth. In the new system, was hell hurtling around in orbit and Dante spinning in his grave?

For the remainder of the 16th century and in the first decade of the 17th, the Catholic Church didn't oppose Copernicus's theories. *De revolutionibus* was read, discussed, and even occasionally taught at Catholic universities, and Catholic scholars used computations based on Copernicus's work to produce the new Gregorian calendar in 1582. On the Protestant front, Luther didn't follow through with any more hostile statements. Calvin observed that the Holy Spirit 'had no intention to teach astronomy'. The anti-Copernican statements often attributed to Calvin are fictitious – a late-19th-century invention. Unlike modern Christian fundamentalists, neither Luther nor Calvin claimed that the Bible was the sole authority in matters other than faith and conduct. Of the ten aforementioned Copernicans between 1543 and 1600, seven were Protestant, three were Catholic. Meanwhile, as the poet John Donne sagely observed, 'most men lived and believed just as they had done before'.

As time passed and the end of the 16th century approached, Copernicus's solutions and techniques and Copernican tables began to seem more and more convincing and indispensable to successive generations of mathematicians and astronomers, and each generation was less wedded than the last to Ptolemaic assumptions. Owen Gingerich, with whom we began this chapter, has made it a sleuthing project to track down existing early editions of *De revolutionibus*, to catalogue them and to see what scholars of the 16th century (in the tradition of 'glossing' that still survived from the Middle Ages) penned in the margins. He has discovered that marginal comments, clarifications and criticisms were passed from teacher to pupil, sometimes for several generations, branching out family-tree fashion. Gingerich tells us in his book *The Great Copernicus Chase*, 'These second-hand annotations reveal that even if Copernicus's revolutionary new doctrine failed to find a place in the regular university curriculum, a network of astronomy professors scrutinized the text and their protégés carefully copied out their remarks, setting the notes on to the margins of fresh copies of the book with a precision impossible by aural transmission alone.' Undoubtedly Copernicus's ideas were spreading and having an increasing impact. It appeared as though Copernican astronomy were headed for a peaceful victory, and scripture could and would be reinterpreted to agree with a new vision of the universe.

In the ancient Hellenistic world, Aristarchus's proposal had failed to overturn Earth-centred astronomy. What was different now that made the change more acceptable? First, while Aristarchus had simply raised the suggestion that the Sun, not the Earth, lay at the centre of the universe, Copernicus gave his readers a good deal of mathematics to chew on, mathematics that they found interesting and useful even if they didn't follow Copernicus all the way. Also, there was strong support from observational astronomy coming along early in the next century – new data that Ptolemaic astronomy of that era had difficulty explaining, and Copernican astronomy explained with ease.

But to understand more fully why an idea that was ignored in the ancient world should finally make its impact 17 centuries later, it's necessary to look beyond astronomy and mathematics and notice that the 16th and early 17th centuries in Europe were an era of intense intellectual, religious, political and cultural ferment. There was a mounting spirit of upheaval and distrust of old assumptions, across the board. Anti-Aristotelianism and humanism were challenging and infiltrating scientific thought with Neoplatonic preferences for geometric and mathematical harmony and simplicity that Ptolemaic astronomy could not provide. Luther and Calvin and their followers, and Henry VIII as well, were calling into question the ultimate ancient authority of the Roman Catholic Church, offering, in its place, not one authority and doctrine but a rich and confusing choice. The Catholic Church was also of many minds on many fronts, so that it is misleading to speak of 'the Church' as though it were a monolith, holding one opinion. The great age of exploration was underway. Columbus had sailed for the New World when Copernicus was in his late teens. Ptolemy's ancient maps were proving to be inaccurate. Was there perhaps reason to distrust his astronomy as well? Ancient copies of Ptolemy's work had turned up and made it impossible to sustain the hope that problems with his astronomy stemmed only from Arabic misinterpretation that could be corrected if scholars knew what Ptolemy had actually said himself. This was a world in many ways ready to entertain the exhilarating thought that with Copernicus humankind had finally not only caught up with, but surpassed, the thinking of the ancients and was ready to move onward, unintimidated by the past.

Nicolaus Copernicus didn't single-handedly overturn the Earth-centred view that had prevailed since ancient times. But he did push open the door that would lead us towards our modern understanding of the universe. This time, the door wouldn't be closed again. As Hoyle puts it, 'It is because Copernicus focused the attention of the world at precisely the right spot, the place where Nature simply had to give up her secrets, that today we judge his work to have been so important.' Hoyle uses a mountaineering term: Copernicus discovered the 'point of attack'. Others would soon be gearing up to make the climb.

CHAPTER 3

Dressing Up the Naked Eye
1564–1642

> I do not feel obliged to believe that the same God who has endowed us
> with sense, reason, and intellect has intended us to forgo their use.
>
> Galileo Galilei

It would be difficult to imagine two educated men of the same historical period more different from one another in background and personality than Johannes Kepler and Galileo Galilei. Kepler was a quiet, introspective man from Weil der Stadt, Württemberg, on the outskirts of the Black Forest near Stuttgart. Galileo was a colourful, feisty, larger-than-life character who hailed from Florence and Pisa when these were world-class centres of wealth, political power and artistic and intellectual ferment.

Kepler's family background was respectable – his grandfather had been a burgomaster – but his father was an evil-tempered ne'er-do-well who abandoned his family; and Kepler's mother was a malicious troublemaker who dabbled in the occult and barely escaped burning as a witch. Galileo's father was a well-educated trader, with a reputation in Florentine intellectual circles as an accomplished musician and music theorist.

Kepler was an unassuming, private man who didn't win friends easily, who struggled for a living teaching, which he didn't do particularly well, and casting horoscopes, which he reputedly did very well indeed, meanwhile pursuing his real passions: mathematics, astronomy and philosophy. Galileo won both friends and enemies readily, relished the spotlight and lived a public, even celebrity life. He thought highly of himself, and he had a flair for self-promotion and an astounding talent as a writer for conveying his scientific ideas to non-expert readers, including some in high places whose favour he curried.

Kepler was a devout Protestant; Galileo a staunch Catholic.

But far more significant than any of these differences was the contrast between the ways in which the two men approached their science. Kepler had a mind that moved by leaps of fancy and intuition and a Neoplatonic Christian faith that the universe, created by God, must have a beautiful hidden harmony to it – that things as far apart as music and geometry and cosmology must have connections, must fall into place and explain one another. Though Galileo made mistakes – insisting, for instance, that the tides were powerful evidence of the correctness of Copernican theory – it is the bulk of Kepler's work that appears to modern eyes much wider of the mark. His most celebrated discoveries seem like small islands of dazzling insight in a sea of wild, woolly thinking. In fact, much of what Kepler wrote appears completely crazy, until we recall that his greatest contributions to our understanding of the universe were at first equally much flights of intuition and imagination, and equally motivated by his longing to uncover hidden harmony, symmetry and relationships. It took a mind that could think as 'far out' as Kepler and follow as many false leads as he did – and then proceed to pin ideas down with conscientious mathematical rigour – to discover connections that really do exist. Galileo, on the other hand, started from what he observed, whether that was a swinging chandelier in the church at Pisa or tiny stars parading around the planet Jupiter. He had a firm belief that the only way to learn the truth about nature was to examine it directly and put it to the test. Though eager and able to speculate about the implications of his discoveries, he was reluctant to espouse publicly or even among friends any ideas for which he didn't see clear support from his own experiments or observations.

For all their genius, each of these men was also remarkably favoured by happenstance. Each had fall into his hands something without which his most celebrated discoveries would never have been made. For Kepler that was the naked-eye astronomical observations of the great Danish astronomer Tycho Brahe; for Galileo, the telescope.

Together, Kepler and Galileo were responsible for the triumph of Copernican astronomy, yet they never met.

Johannes Kepler was born in 1571, 28 years after the death of Copernicus and the publication of *De revolutionibus*. His boyhood in Weil der Stadt, with his sinister mother and, occasionally, his irresponsible father – in a household full of other unsavoury, unhappy relatives – was as dreadful as anything ever dreamt of by Charles Dickens. The young Kepler was seldom in good health and a childhood illness left him with weak eyesight. Nevertheless, it became clear early on that he had extraordinary intellectual gifts and a deeply religious nature, neither of which endeared him to his schoolmates. He wasn't good at making friends.

Kepler hoped to become a Lutheran minister, and he went on from the local school to a theological academy for the children of 'poor and pious people' (Kepler's family met the first requirement, if not the second) and then to the university at Tübingen, where scholarships paid his way. The Senate of the University observed that Kepler had 'such a superior and magnificent mind that something special may be expected of him'. Tübingen still taught Ptolemaic astronomy, but it was during this period that Kepler became a Copernican, perhaps due to the influence of the private views of the astronomer Michael Mästlin, who was one of his teachers. For an example of what astronomy was like before telescopes: Mästlin's study of the nova of 1572, using no instrument at all except for a piece of thread, gave results more accurate than anyone else's, including Tycho Brahe.

In 1594, when Kepler was 23 and in his final year as a theology student, a teaching job in mathematics and astronomy opened up at a seminary in Graz, Austria. The University of Tübingen nominated Kepler. Though surprised and somewhat distressed at this sudden change in his career trajectory, Kepler accepted the position.

Kepler wasn't a good teacher, but his new job evidently did leave him time to pursue his science and philosophy, for only two years after he began teaching he published the first book since *De revolutionibus* to defend Copernican theory, and he made a far stronger case for it than Copernicus had been able to do. The 24-word title of Kepler's book is usually shortened to *Mysterium* or, translated, *Cosmographic Mystery*. *Mysterium* is a book that elicits, at best, tolerant smiles from modern readers, in spite of its support of Copernicus and the ingenious nature of its proposal. Kepler attempted to explain both the number of planets and the sizes of their orbits in terms of a relationship between the planetary spheres and the five regular solids of geometry: cube, tetrahedron, dodecahedron, icosahedron and octahedron. See Figure 3.1.

The publication of *Mysterium* led to the first known contact between Kepler and Galileo. At Padua, where he was teaching at the time, Galileo received a copy of Kepler's book. He wrote to Kepler saying he was looking forward to reading it and also that he had long believed in the Copernican theory himself but had not said so openly, to avoid ridicule. In a return letter, Kepler urged Galileo to make his opinion public. Galileo didn't take that advice.

Kepler continued eking out a living in Graz as an ineffective

Figure 3.1

(a) The five regular or 'cosmic' solids. In every case, all of the sides are identical and only equilateral figures are used for them. These are the only possible regular or 'cosmic' solids.

(b) Kepler's arrangement of them in relation to the spheres of the planets. Saturn's sphere is outside the cube. Jupiter's sphere is between the cube and the tetrahedron. Mars's sphere is between the tetrahedron and the dodecahedron. Earth's sphere is between the dodecahedron and the icosahedron. Venus's sphere is between the icosahedron and the octahedron. Mercury's sphere is within the octahedron.

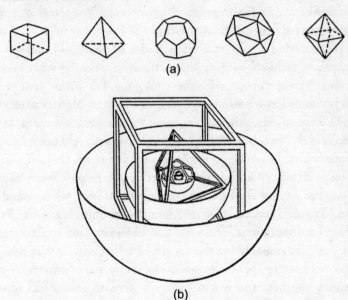

(a)

(b)

teacher, went on with his astronomical work, wrote some books about astrology, and married – a marriage that lasted, though there is some indication it was not an entirely happy one. Then in 1598, the Archduke Ferdinand began to make life miserable for Lutheran leaders and teachers. Things went from bad to worse and, eventually, given a day's notice to leave Graz altogether or be sentenced to death, Kepler departed. It soon became evident that the Archduke's attitude was intransigent. Kepler would no longer be able to live and work in Graz.

However, Kepler was not completely out in the cold, for the opportunity had arisen to join Tycho Brahe at Prague. Kepler had sent the older man a copy of *Mysterium,* and Tycho had recognized a serious talent. Kepler was understandably apprehensive about how well the two of them would get along, for Tycho had a reputation as a proud, imperious eccentric. His nose had been partly sliced off in a duel, and he had restored the missing bit himself with gold, silver and wax. But Tycho was also unarguably the greatest astronomer of his generation . . . and Kepler needed a job. So Kepler, now 30 years old, moved to Prague in 1601. Whatever the difficulties working with Tycho, he didn't have to cope with them for long, for within two years Tycho died. Kepler succeeded him as Imperial Mathematician.

This new position was a considerable improvement over teaching at Graz, but there was a negative side. Although he had Tycho's impressive title, Kepler's salary was much lower than Tycho's had been and often it wasn't paid. Kepler was obliged to waste a great deal of time trying to collect what was due him. The job as Imperial Mathematician did, however, have other compensations, for, over the objections of Tycho's relatives and fortunately for the future of astronomy, it was Kepler who fell heir to Tycho's magnificent set of astronomical observations, the best the world had ever known. Tycho had found that circular orbits were difficult to reconcile with the actual paths of the planets, and he had undertaken an exhaustive series of observations that he hoped would throw more light on the problem. No man was better suited than Kepler to put this precious inheritance to optimum use.

Kepler brought to this work both a firm belief in Copernican astronomy and an unwillingness to accept that the skilled and meticulous Tycho's data could be faulty. The two must somehow fit, even though Tycho himself had rejected Copernican astronomy in favour of the scheme that had the Sun orbiting the Earth and all the other planets orbiting the Sun. Kepler also brought to his task a concept of his own that the Sun moves the planets by a 'whirling force' centred in itself. Orbits centred elsewhere than on the Sun would not do.

Tycho Brahe had paid particular attention to the planet Mars, and it was in trying to make sense of these observations that Kepler finally found himself obliged to abandon circular orbits, removing the stumbling block that had hobbled Copernican astronomy since its inception. He realized that by using elliptical orbits he could explain Tycho's observations. Kepler signalled his overwhelming joy and astonishment at his insight by falling to his knees and exclaiming, 'My God, I am thinking Thy thoughts after Thee.' Those who describe Kepler as dry and passionless and his life as uniformly drab and sad simply fail to appreciate the sorts of things that moved him.

Kepler's discovery and Tycho's earlier observations (without a telescope) that made it possible were achievements that still inspire awe in modern astronomers. The ellipse in which the Earth orbits is so nearly circular that any attempt to make it obviously an ellipse in a scale drawing ends up being a distortion. It is only slightly more obvious that the orbit of Mars is an ellipse. Yet with this seemingly trivial geometric alteration, the Copernican system fell into place. Kepler, aware of the power of his discovery and also of the reaction it would inevitably receive, commented wryly that with this introduction of elliptical orbits, he had 'laid an enormous egg'. He had indeed, and even many who admired his work found this egg difficult to digest. Galileo, for one, never accepted it.

Saying an ellipse is an 'egg shape' or an 'oval' isn't precise enough. It's best to think of an ellipse in one of two ways. One is as a slice out of a cone. Not just any oval or egg shape can be produced by slicing a cone, and so not all ovals and egg shapes are ellipses. A circle, however, *is* an ellipse. It's a slice directly across the cone. See Figure 3.2.

Figure 3.2

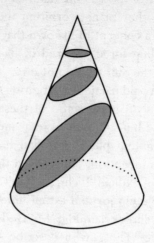

A second way of thinking about an ellipse is a little more difficult to describe, but it gets the same result: imagine a piece of thick cardboard lying on the table before you. Stick two drawing pins into it, a distance apart. Then take a piece of string that is longer than the distance between the two pins and attach its ends to the pins – something like the string shown in Figure 3.3.

Figure 3.3

Next take a pencil and pull the string taut with its point (Figure 3.4), still keeping the string flat on the surface of the cardboard.

70

Figure 3.4

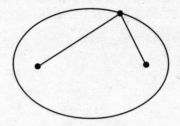

If you move the pencil, allowing it to slide along the string while the string remains taut, the point of the pencil will draw an ellipse. Changing the length of the string or moving the drawing pins closer together or further apart changes the shape of the ellipse, just as changing the tilt of the slices in the cone does. The closer together the two drawing pins are, the nearer the ellipse comes to being a circle. If they are in precisely the same location, it *is* a circle.

Each drawing pin represents a *focus* of the ellipse, so an ellipse has two foci. If a planet's orbit is near to circular, that means that the ellipse's foci must be very close together. By contrast, comets, which also orbit the Sun in elliptical orbits, have foci that are very far apart, producing an extremely elongated ellipse.

The first of Kepler's two laws of planetary motion that appeared in his book titled *Astronomia Nova* (*New Astronomy*) states that a planet moves in an elliptical orbit, and the Sun is located at one of the two foci of that ellipse (see Figure 3.5).

Kepler's second law has to do with the variation in the speed of a planet as it travels in its orbit. The law states that an imaginary straight line (called the 'radius vector') joining the centre of the planet to the centre of the Sun 'sweeps out' equal areas in equal intervals of time.

Picture an elliptical orbit, the Sun (one of its foci), and a planet travelling along in the orbit (see Figure 3.6). Picture the

Figure 3.5

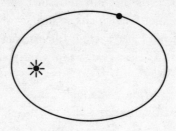

planet at **A** and draw an imaginary straight line from **A** to the centre of the Sun. Set a stopwatch and time the planet as it orbits to **B** (which for purposes of this demonstration could be located anywhere on the part of the orbit nearer the Sun). Think of the imaginary line moving with it like the hand of a clock, shading the area it sweeps across. At **B**, stop the shading and the timing. You have now created a shaded area that we say is 'swept out' as the planet travels from **A** to **B** – looking like the left side of Figure 3.6. You know how long it took the planet to travel that distance and sweep out that area. Wait until the planet has travelled somewhat further. Decide on a point **C** somewhere on the side of the ellipse further from the Sun.

Figure 3.6

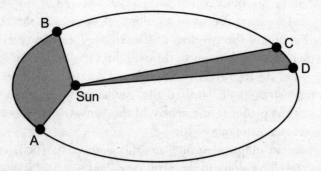

When the planet reaches that point, again draw a line from the planet to the centre of the Sun and start the shading and the stopwatch. The planet travels along and so does the sweeping line. When the planet has travelled for *the same amount of time it took to go from* **A** *to* **B**, stop the shading and call that final point **D**.

The resulting drawing resembles Figure 3.6. What does it mean? Notice that on the left side of the picture, from **A** to **B**, the planet travelled a much greater distance, in the same interval of time, than it did from **C** to **D** over on the right side. Clearly it couldn't have maintained a constant speed. It had to travel much faster from **A** to **B**, to cover all that distance, than it did to cover the smaller distance from **C** to **D** in the same interval of time. The two shaded 'swept out' portions, however, are equal in *area*. Kepler's law thus predicts that a planet will always travel at a faster rate the nearer its path comes to the Sun. Kepler thought the variation in speed occurred because the distance from the Sun affected how much or how little the planet felt the Sun's 'whirling force'.

Kepler's second law, then, accounted for the way a planet varies in speed and made it possible to predict those variations. While one can't make sense of the system by saying 'this planet always travels thus-and-so-many miles per hour', one *can* make sense of it by saying 'this planet's imaginary line to the Sun always sweeps out an area of thus-and-so-many square miles per hour'. This was another of Kepler's crazy ideas, a 'connection' he found between astronomy and geometry, arising from his belief in the intrinsic harmony of a universe created by God. We still today think he got this one right.

Kepler's *Astronomia Nova* contained more than these two laws. He was well on track towards our modern understanding of gravity, previewing Newton's later discoveries about the tides and understanding them a great deal better than Galileo did. Kepler had discerned that the more massive a body, the stronger its attractive effect. He also wrote that it couldn't be the Sun's light that provided the 'whirling force', for the Earth didn't screech to a halt during a solar eclipse.

Kepler finished *Astronomia Nova* in 1606 but didn't publish it until 1609. That was the same year Galileo first looked through a telescope.

In 1611, after the death of Kepler's wife and son, and with difficulties for Protestants mounting in Prague with the counter-reformation as they had done earlier in Graz, Kepler moved to Linz. He lived there for 14 years, marrying a second time. It was in Linz that he published, in 1619, a book called *Harmonices Mundi* (*Harmonies of the World*), relating the orbital speeds of the planets to melodic lines. He speculated that it might be possible to discover the moment of Creation by calculating backwards in time and determining the moment when the orbits of the planets would have produced the equivalent of the most perfect musical harmony.

Much more significant, *Harmonices Mundi* also contained Kepler's third law of planetary motion, which establishes a relationship between the lengths of time the planets take to complete their orbits and their distances from the Sun. Those not conversant in the language of mathematical equations may skip the following paragraph.

The equation is $(T_A / T_B)^2 = (R_A / R_B)^3$. T_A is the time it takes planet A to complete its orbit (its 'orbital period'). T_B is the time it takes planet B to complete its orbit. R_A is the average distance from planet A to the Sun. R_B is the average distance from planet B to the Sun. The ratio of the squares of the orbital periods of the two planets is equal to the ratio of the cubes of their average distances from the Sun.

This equation makes it possible to build a scale model of the solar system. The question remains, however: What is the scale? For even though the orbital periods could be timed and were known, Kepler's law didn't provide the means to calculate any *actual measurement* for the distance from any planet to the Sun.

The third law was an equation waiting for just one absolute, known distance to plug in.

Kepler's discovery of his first and second laws allowed him to put together far more accurate tables for calculating the planets' positions for any time in the past or future. When he published his third law, he had already begun the laborious task of producing these, but the completed tables were still slow in coming. Kepler's preface explains that this was partly because of difficulties finding financial backing (he finally paid for the publication out of his own pocket) and also because of 'the novelty of my discoveries and the unexpected transfer of the whole of astronomy from fictitious circles to natural causes'. He went on to point out that no one had ever attempted anything of this kind before. It required intense study of Tycho's astronomical observations. Kepler tried orbit after orbit, leaving one after another behind when, after prodigious computation, they didn't match his data. He was not particularly happy doing this work. It must have been very pedantic for so imaginative and creative a mind. It wasn't until 1627 that he finally published his *Rudolphine Tables*, and they were a triumph for Copernican astronomy.

Kepler spent his last years at Sagan, in Silesia. He died at Regensburg on a trip seeking a new job and trying to collect back salary, in November 1630, just a year short of seeing that his prediction, in accordance with the Rudolphine tables, that Mercury would cross the disc of the Sun on 7 November 1631 was correct. He had composed his own epitaph:

I measured the skies, now the shadows I measure
Skybound was the mind, Earthbound the body rests.

★　★　★

Compared with Kepler's bleak childhood, Galileo's apparently was a pleasant one in a family that valued intellectual pursuits. He was born seven years before Kepler, in 1564, in the north Italian city of Pisa, and attended school both there and in

Florence. At the age of 17, honouring his father's wishes, he entered the University of Pisa as a medical student. Two years later, when it was clear that mathematics and mechanics, not medicine, were Galileo's forte, his father allowed him to change his course of study.

One of Galileo's important discoveries took place during his student years. Good Catholic he may have been, but evidently he was not always completely attentive at services. A lamp swinging on a long cord in the cathedral caught his eye. He noticed that regardless of whether the length (in space) of each swing was long or short, it seemed that the time it took to complete the swing remained the same. Curious, he experimented on his own and found that the length of time it takes a pendulum to complete one swing depends not on how big the swing is but only on the length of the cord by which it hangs.

In 1585, Galileo's formal education ended when his father ran short of money to pay university costs. Galileo came home to Florence, where his family was now living. Undeterred, he studied on his own and with scholars among his father's acquaintances and was soon making a local reputation with several inventions and discoveries, circulating his own short book on measuring the specific gravities of bodies, and brashly voicing suspicions about Aristotle's mental capacities. One bizarre undertaking during these years was a series of public lectures he delivered about the shape, size and location of Dante's hell — a topic that might draw a considerable audience even today.

In 1589, when Galileo was 25, four years after he had left the University of Pisa without a degree, he returned there as a lecturer. Pisa, like Tübingen, still taught Ptolemaic astronomy. Whatever Galileo may have been thinking privately, that was what he taught. It isn't clear just when he became a convinced Copernican.

A discovery Galileo made while he was a lecturer at Pisa caused his already low opinion of Aristotle to sink even further.

Though Aristotle had in truth been rather ambiguous on the subject, at least in those writings that survive, most scholars of Galileo's time, including Galileo, thought that Aristotle had said that if two objects were dropped simultaneously from the same height, the heavier would strike the ground first. Galileo either dropped weights on numerous occasions from the Tower of Pisa, as his earliest biographer reported, or rolled them down a ramp, or perhaps he tried both. But he established to his satisfaction that the heavier and the lighter hit the ground at the same time. This was, of course, an experiment that he couldn't perform in the absence of air resistance, and science historian Thomas Kuhn has quipped that it probably wasn't Galileo who carried out the legendary public demonstration from the Leaning Tower but a defender of Aristotle, who thereby proved quite decisively for all present that Galileo was wrong and Aristotle was right. In the 20th century, astronauts performed the experiment in the airless environment of the Moon. Galileo was right.

Never a tactful man, Galileo somewhat incautiously proceeded to debunk the still highly venerated Aristotle. Aristotle 'wrote the opposite of truth' . . . was 'ignorant'. Galileo also showed a lack of judgement when he bluntly advised the Grand Duke of Tuscany, Ferdinand I (who had granted him his professorship), that a dredging machine designed by the Grand Duke's brother-in-law wasn't going to work. The machine was built nevertheless and *didn't* work, which made Galileo even less popular. When his father died in 1591, leaving him as eldest son responsible for the family, Galileo's salary at Pisa was far from sufficient and, thanks to the dredging machine incident, not likely to rise. He looked for a new job and found one at the University of Padua, where Copernicus had studied medicine 90 years earlier. The Venetian Republic, the source of funding for Padua, was liberal and tolerant compared with Tuscany, and Padua and Venice were much more sympathetic venues for someone with ideas as radical and speech as imprudent as Galileo's.

Though his financial difficulties didn't end, Galileo was happy and influential in the intellectual milieu of Padua and Venice. On the personal front, he fathered three children by Marina Gamba, though he never married her. Their relationship seems to have ended amicably when he finally moved back to Florence, for he remained good friends with her and with the man she later married. Galileo's two daughters both became nuns, and one of them was a source of great comfort and help to him when he was elderly.

At Padua, Galileo still taught Ptolemaic astronomy. However, though he lacked observational evidence to support a change of allegiance – and he was almost always careful about waiting for that before making announcements or going into print – by the mid-1590s he was personally convinced that Copernican astronomy was correct. In 1597 (12 years before he first looked through a telescope) he wrote a letter to that effect to a friend at Pisa, and it was also at this time that the exchange of letters with Kepler took place concerning Kepler's book *Mysterium*. Galileo had joined the Copernican camp, but he declined to say so publicly.

In the late spring of 1609, the same year that Kepler published *Astronomia Nova* with his first and second laws, Galileo, in Padua, heard of an invention from Holland currently on view nearby in Venice – a tube with lenses arranged in it so that it made objects in the distance look closer. Apparently Galileo didn't hurry off to examine this wonder in person. A few days later he heard a report of another such instrument from a friend in Paris. His interest aroused, Galileo started pondering what arrangement of lenses would produce the reputed effect. Venice and its nearby islands were centres of expert glass-making, so there was no difficulty obtaining the lenses he needed to build his own improved 'perspicillum'.

The use of lenses for eye-glasses was nothing new. That had begun as early as the 13th century. There were probably also telescopic devices before the 17th century. However, one of the first pieces of documentable evidence of such an instrument identifies the maker as Jän Lippershey, a lens-grinder from the Dutch island of Walcheren. He presented it to Dutch authorities in October of 1608, eight or nine months before it came to Galileo's attention.

It was obvious that such an instrument would be useful for sighting ships and distant features of the landscape. One brochure in the autumn of 1608 also pointed out its advantage for 'seeing stars which are not ordinarily in view, because of their smallness'. Sir William Lower, a Welshman, looked at the Moon through a telescope earlier than Galileo did and thought it looked similar to a tart: 'here some bright stuff, there some dark, and so confusedly all over'.

Clearly, the familiar story that Galileo invented the telescope is untrue. By the time he knew of its existence it had already been on sale in Paris and probably elsewhere for several months. A second piece of fiction is that he tried to pass it off as his own invention. However, Galileo did proceed immediately to make better capital out of it than anyone else was doing. On this occasion, and perhaps several others (it isn't always clear precisely where he got his ideas), Galileo displayed a talent for seeing the unrealized potential of another person's thought or invention and carrying it forward so rapidly and enthusiastically that he was halfway over the horizon before its originator had left the starting line. That isn't plagiarism, but it did, in the case of the telescope at least, result in Galileo getting the popular credit while (his letters show) he was actually giving fair credit to others.

Galileo, a master of self-promotion, presented his own improved version of the tube with lenses to the Senate in Venice, hustled some of them up to the top of the Campanile, and showed them that it was possible to look out to sea and spot ships that wouldn't be visible to the naked eye until two hours later. The military and commercial advantages of such an instrument were obvious to rulers of a major city-state whose prosperity rested on trade by sea. Galileo received a permanent appointment at the University of Padua and a hefty increase in salary.

But Galileo had uses in mind for his 'perspicillum' other than spotting ships, providing curious occasional glimpses of the Moon and stars, and securing university tenure at Padua. He set about making systematic astronomical observations, recording them, and using his fine mathematical skills to draw conclusions about what they meant.

In the autumn of 1609, Galileo turned one of his new instruments on the Moon. Although the ancient Greeks had described the Moon as 'earthy, with mountains and valleys', conventional wisdom in Galileo's day had it that it was perfectly smooth and spherical. Both Greek and Hellenistic astronomers and medieval scholars understood that the Moon shines by reflected sunlight, not by its own light. Through his telescope Galileo watched sunrise on the Moon's surface and saw isolated bright dots in the dark portion expand and join with one another. Reminded of what he had observed when sunrise strikes mountain peaks on Earth, he speculated that the separate bright spots must be peaks and ridges, lit first by the Sun's rays before these could penetrate to the lower areas of the Moon's surface. It occurred to him that by studying the shadows of these features he might measure the heights of the peaks and ridges. Galileo arrived at an estimated height of four to five miles. Modern measurement of the particular range of lunar mountains that he studied has them no higher than 18,000 feet. Never mind that discrepancy. The more significant point was that the Moon was not, as many had supposed, smooth.

Aiming his telescope at things more distant than the Moon, Galileo began to make further discoveries. In January of 1610, using an instrument whose lenses he had ground himself with great care, he discovered three pinpricks of light near Jupiter, neatly lined up with the planet. Galileo watched, mystified and then increasingly excited, as the little stars and Jupiter exchanged positions in their line-up and varied in brightness over the course of several nights. See Figure 3.7.

Galileo concluded that this remarkable heavenly quadrille 'ought to be observed henceforward with more attention and precision'. Before long he found that there are four rather than three stars; that the stars move within a narrow range, always in line with Jupiter and with one another; that they stay with the planet when its motion becomes retrograde; that when they are furthest from Jupiter they are never closely packed together, but when they are nearer to Jupiter they are sometimes closely packed. The implication of this last was that if the stars are circling Jupiter, the orbits in which they move are not all the same. If the stars were following one another in the same track, it's likely they would sometimes line up so as to seem (from our vantage point) to cluster when they are furthest from the star.

Galileo reasoned that these could only be satellites, 'planets never seen from the beginning of the world up to our own time', orbiting Jupiter in the same way the Moon orbits the Earth. The deeper importance of what he had discovered also didn't escape him. Never again would it be possible to suppose that there was only one body that was the centre of all motion in the universe.

Galileo, being Galileo, soon found a way to capitalize on his discovery. He rushed into print with a book called *Sidereus Nuncius*, translated *Starry Message* (the quotations in Figure 3.7 come from that book), calling on all astronomers to equip themselves with good instruments and turn them on Jupiter. He dedicated his book not to just any local nobleman, but to the powerful Grand Duke Cosimo II de' Medici, of Tuscany, who had once been his pupil. He decided to name his discovery the Cosmican Stars – in honour of Cosimo – but soon thought better of that. It would sound too much like 'cosmic' and the significance of the name would be missed. He settled on the Medicean Stars. There were, after all, four stars and four Medici brothers.

Figure 3.7

7 January 1610
(The large disc in this picture is Jupiter. Galileo assumed the three little 'stars' were part of the background of fixed, distant stars – though they made him 'somewhat wonder'.)

8 January 1610
(Jupiter seemed to have passed up the three stars, and Galileo 'became afraid lest the planet might have moved differently from the calculation of astronomers'.)

9 January 1610
Clouds

10 January 1610
(Galileo decided the third star must be hidden by the planet and it also occurred to him, 'changing from doubt to surprise – that the interchange of position belonged not to Jupiter but to the stars'.)

(Galileo waited 'with the most intense longing'.)

11 January 1610
(Galileo noticed that one of the two visible stars was larger than it had been before and quite a bit larger than the other. He 'decided unhesitatingly, that there are three stars in the heavens moving around Jupiter'.)

Galileo meanwhile hadn't been neglecting other stars and planets and had found that, though his instrument transformed the planets into discs, the stars still looked like points of light. Furthermore there were astounding numbers of them that had never been seen before. The Milky Way, he discovered, is 'nothing else but a collection of innumerable stars . . . many of them are tolerably large and extremely bright, but the number of smaller ones is quite beyond determination'. There was far, far more to the universe than anyone on the face of the Earth had ever supposed. Galileo hastily added some pages in the middle of his book to report these discoveries.

A copy of *Sidereus Nuncius* reached Kepler in Prague. He also heard about the discovery of Jupiter's satellites through his friend Wackher von Wackenfels. Galileo asked Kepler for his opinion and the reply came in the form of a long letter that was later published as *Conversation with the Starry Messenger*. In it, Kepler discussed Galileo's discoveries and theories and expressed his agreement. Galileo wrote back, 'I thank you because you were the first one, and practically the only one, to have complete faith in my assertions.' Galileo did not, however, respond to Kepler's rather broad hint that he would enjoy owning one of Galileo's telescopes, though he was sending them as gifts to many influential people. Galileo actually understood the principles of the telescope no better than Kepler, perhaps not as well, though Kepler never built one. When Kepler was working with Tycho's data he had to study optics to learn how to eliminate errors due to the smearing ('refraction') of light as it passes through the Earth's atmosphere. Kepler did, for a short while, have one of Galileo's telescopes on loan from a mutual acquaintance.

Galileo's self-marketing scheme was successful. His book appeared in March 1610 – quick publishing indeed – and by late summer he had accepted the Grand Duke's offer of a job and moved back to Florence. He was now a celebrated and well-placed scientist and astronomer.

About the time Galileo must have been unpacking his equipment in Florence, the planet Venus came into a good position for viewing in the evening sky. Galileo examined the planet and the area around it, searching in vain for companions like the ones he had found around Jupiter. In a letter in mid-November to Cosimo's brother Giuliano, ambassador in Prague, he wrote that there seemed to be no satellites around any of the planets except Jupiter. However, his study of Venus was to yield other significant results. Although most scholars give the credit to Galileo, there is some question whether it was he or his former student Benedetto Castelli who at this juncture remembered a suggestion that Copernicus had made in *De revolutionibus*, that Venus might supply important evidence in the case against an Earth-centred universe. It was certainly Galileo who proceeded to find the evidence.

When we visited the amusement park in Chapter 2 and studied the moving lights, there was one possibility we failed to consider. Imagine once again the entire park plunged into darkness, with glowing lights attached to the heads of only a few of the riders. As we try to figure out what carnival rides might produce this pattern of movement – and what the park would look like in daylight – we should bear in mind that some of the pinpoints of light we see might not be light sources at all, but instead be shining by reflected light. Perhaps a bauble that is not a light source itself, on the head of one of the carousel horses, is reflecting the light cast from a nearby horseman. How would we know the difference?

Do any of the lights have 'phases' like the Moon? Do any of them sometimes appear as a distorted disc, a half, or a crescent? If we find that, might it indicate that, like the Moon, this is not a light source but a reflection of light coming from elsewhere? If so, perhaps we could use its 'waxing and waning' as a clue to its position and motion, and to the position and motion of the source of its light.

It was this line of reasoning that Galileo used in 1610, when

Figure 3.8

All three systems are able to explain the positions of Venus, but the Ptolemaic system cannot explain the phases.

In the Ptolemaic system, with Venus always between the Earth and the Sun – travelling on an epicycle on a deferent with the Earth as its centre – an observer on Earth would never see the face of Venus anywhere near fully illuminated. This figure shows Venus in three positions.

In the Copernican system, below, in which both Venus and the Earth orbit the Sun, Venus has almost a full set of phases, as the Moon does.

In Tycho Brahe's system, in which Venus orbits the Sun while the Sun orbits the Earth, Venus has the same set of phases it has in the Copernican system, left.

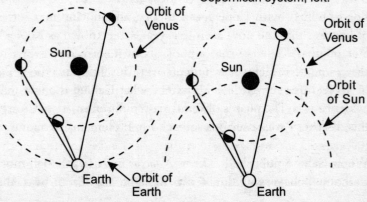

he studied the planet Venus through his telescope. In Ptolemaic astronomy, Venus always lay between the Earth and the Sun. For that reason, *if* Venus sheds no light of its own but only shines with reflected sunlight, observers on the Earth should never see the face of Venus anywhere near fully lit. In other words, it should never be the near equivalent of a full Moon. See Figure 3.8.

From August till October of 1610, Venus would have appeared as a blurry disc through Galileo's telescope. In October he would have seen the disc flatten to a lozenge. Galileo knew then that Venus was shining by reflected light from the Sun, not by its own light. From November till January, Venus would have waned to a crescent in the same manner the Moon does. Galileo was aware that in the Ptolemaic system it would have been impossible for Venus to have nearly a *full range* of phases, even if its epicycle had been miscalculated and was actually on the other side of the Sun from Earth. To put it bluntly, Galileo could not have seen what he saw if Ptolemaic astronomy had been correct. In the Copernican system, and in Tycho Brahe's system (let us hasten to admit), it was what one would expect to see.

Finally Galileo had found persuasive observational evidence that Ptolemaic astronomy was inferior to Copernican astronomy. To Galileo's mind this was actually not the first evidence he had found. Six years earlier, in 1604, he had first sided publicly with Copernican theory, announcing in a series of lectures that the nova seen that year, later known as Kepler's Star because Kepler wrote a book about it, provided evidence that some of Ptolemy's arguments were invalid. Since no texts of his lectures survive, it's a mystery what Galileo thought that evidence was. The phases of Venus are a different matter. Clearly this discovery was a serious setback for Ptolemaic astronomy.

When Galileo published *Sidereus Nuncius* in 1610, though most Catholic believers undoubtedly thought the Earth was the centre of the universe and assumed that scripture supported this view, the Catholic Church had no official policy regarding the arrangement of the cosmos. Contrary to modern popular legend, it had succeeded in staying clear of the debate ever since the appearance of *De revolutionibus*, continuing a centuries-old practice of tolerating diversity when it came to cosmological arguments. Nicholas of Cusa had proposed a moving Earth before Copernicus, and he was a cardinal and papal legate. The Church hadn't criticized or condemned him. Giordano Bruno, a scholar strongly influenced by Cusa, had been burned at the stake primarily because of his heretical religious views, not because of his scientific ideas – though those didn't help. Two of Copernicus's strongest supporters had been powerful Church officials. Now, in the wake of Galileo's discoveries, many among the Catholic hierarchy appear to have been hoping the Church could continue to make no official pronouncement on this matter, and some among them were particularly anxious that it should not take a fundamentalist stand on the interpretation of scriptural passages having to do with cosmology. The quote often attributed to Galileo – 'Scripture teaches how to go to heaven, not how the heavens go' – actually was not from him but from one Cardinal Baronis. The general, vague state of truce seemed to be that if all parties could avoid saying that any scientific arrangement of the universe *or* scriptural cosmological statements should, *or should not*, be treated as literal truth, no one would be taken to task and everything would continue to go smoothly. That was a truce that was violated on all sides but nevertheless continued to prevail for a time.

As news of Galileo's discoveries spread, the most outspoken and dogmatic reactions came from conservative astronomers in the universities, who still continued to lisp rote 'truths', insisting that the authority of Aristotle and Ptolemy must not be questioned. Why bother to observe nature or look through a telescope when Aristotle and Ptolemy had already given the answers? Galileo's battle with these men was not only over the arrangement of the universe. His way of doing science, his insistence on examining and testing nature to learn about it, was to these intransigent scholars foolishness at best, scientific heresy at worst.

However, not all who opposed Galileo were so lacking in intelligence and valid arguments. Even modern astronomers and historians of science admit that Galileo's case for Copernican astronomy was not as open and shut as he insisted it was. Some of Galileo's contemporaries argued, correctly, that his 'evidence' was not 'proof'. Jesuit scholars pointed out that while Copernican theory was capable of explaining Galileo's discoveries, Tycho's theory could explain them equally as well without changing the centre of the universe. Many astronomers, picking up on what they thought was Copernicus's preface to *De revolutionibus*, were willing to accept the Copernican arrangement as an excellent hypothetical model that 'saved the appearances', while not making decisions whether or not it actually represented reality.

Galileo himself did not have a great many personal supporters, and his scientific views were not the only reason. He had never learnt to curb his arrogant tongue. He was careless of fragile egos and didn't suffer even the most reasoned opposition graciously. In fact he had a tendency to think of anyone who disagreed with him as an enemy. Vitriolic, insulting statements that he made in letters and in person didn't make him popular among his colleagues.

Galileo actually bears a large share of the blame for making the Church the arena where the most visible clash between the two theories took place. Remarkably, given the Galileo 'legend', he didn't at all oppose the idea that the Church should exercise authority in scientific matters. Quite the contrary: he was impatient and scornful of its reluctance to face up to the issue. Modern scholars would consider it healthy that the Church had no official position about the design of the universe. Galileo did not. What was important to him was that Church authority weigh in on *his* side. He was supremely confident that that would happen.

Though Church officialdom continued to be stubborn about involving itself publicly in this squabble, behind closed doors in the upper echelons of power it seems reasonably certain that the question was not being ignored and that it was considered a delicate matter. Several intellectuals among the Church leaders had concluded that Sun-centred astronomy was preferable to Earth-centred astronomy. If that was the case, the Holy Office would be wise to support Earth-centred astronomy; it eventually would have to. Some thought it best to continue the hands-off policy and let matters drift in that direction, if they did, on their own. Nearly all agreed that the Church should proceed slowly and with caution. The Catholic Church had long felt a duty to protect its members, particularly the less well-educated among them, from ideas that threatened to undermine simple faith, however much that simple faith might seem *too* simple to more sophisticated Catholics. The belief systems of most parishioners evidently included the Aristotelian/Ptolemaic arrangement of the universe, though how devoutly attached to that they were is difficult to judge. Arguably this was one reason a gradual shift seemed far preferable to any sudden move. It was much less likely to shock and confuse. But also – quite apart from any concern it might have had for the spiritual well-being of its members – given the tenor of the times and the ongoing struggle with Protestantism, Church leaders could ill afford to risk any major unsettling of the flock.

In this setting, Galileo was the proverbial bull in the china shop. He had too much zeal for his cause to recognize that forcing the issue was not in his own best interest. He was impatient. He knew that powerful people were agreeing with him. Surely others could not fail to see that on both scientific and scriptural grounds, Copernican theory was correct and acceptable.

In 1611, Galileo paid a visit to Rome and was received with

enormous respect and friendliness. But two years later he made a serious error. He wrote letters to his former student Castelli and to the mother of the Duke of Tuscany, stating that Copernicanism should be treated as fact and commenting that it would be unwise to interpret passages of scripture in such a way as to force them to support interpretations of nature that might later prove obviously wrong. (It seems Galileo thought that only a 'forced' interpretation of scripture could support Ptolemaic astronomy.) Some in high Church circles had been voicing similar sentiments, recognizing that this was not a contest between literal and metaphoric reading but a matter of *which* metaphoric reading. More damagingly, Galileo also said in his letters that scripture had not gone into more sophisticated scientific detail which *could* be taken literally because it was written to be understood by 'common people who are rude and ignorant'.

Galileo's letter got passed around and soon some were interpreting it (or deliberately misinterpreting it) to say that Galileo was questioning the validity of scripture. When pressed, a committee of the Holy Office looked into the matter and ruled that Galileo's letter was *not* heretical. However, he received a letter from the powerful Jesuit Cardinal Roberto Bellarmine, advising him that until there was definite proof that the Earth moved, that idea ought to be treated as hypothetical and scripture interpreted in the more commonly accepted way. Bellarmine was one of those who believed that that was how Copernicus himself had regarded his model.

In 1616, Galileo was once again in Rome pushing his campaign for Copernicanism. This time the Holy Office, increasingly beleaguered by Galileo's enemies, sent a survey of sorts to leading university scholars regarding the centrality of the Sun and the motion of the Earth. It isn't known who drew up the mailing list, but the replies were overwhelmingly anti-Galileo: he was scientifically wrong and philosophically heretical.

Galileo's opponents won a qualified victory in this round. The diary of one Giovanfrancesco Buonamici, a diplomat from Tuscany, states that two cardinals opposed the Pope's inclination to declare Copernicanism contrary to the faith. Bellarmine was one of them. The other was Cardinal Maffeo Barberini, a good friend of Galileo. In the end the judgement was that Copernicanism was 'contrary to Holy Scripture and cannot be defended or held'. However, that judgement was not given the stamp of papal authority, and this meant that Copernicanism was not officially heresy. That seems a trivial legal distinction, as does another, that Galileo had been *informed* of the Church's decision and *admonished* to abandon Copernican views until he had unassailable proof, but *not* punished or forbidden to teach Copernicanism. However, Galileo recognized this last point as so crucial that he asked for and received a letter from Bellarmine, who had delivered the Church's verdict to him, making the details clear. Galileo did leave off campaigning for Copernican theory for a while and turned to other work, biding his time and watching for a better political moment to resume that effort.

In 1623, that moment seemed to have arrived. Liberal Catholics rejoiced as Cardinal Barberini became the new Pope Urban VIII. Barberini and Galileo had feasted often together and enjoyed discussing science and philosophy. Barberini, as a cardinal, had successfully opposed the papal decree that would have declared Copernicanism heretical. Some accounts have it that now, in 1623, Barberini told Galileo that as Pope he could no longer chat off the record with Galileo and express his own opinions freely. Galileo asked to be allowed to write a book on *both* systems, and he took Barberini's response, whatever that actually was, as encouragement.

By this time, in spite of the 1616 stalemate in Italy, Copernican astronomy – with Kepler's elliptical orbits and his three laws, Galileo's discoveries, and the Rudolphine tables soon to be published – was very near to holding its winning hand. But if

Galileo thought the Church was ready to put its cards on the table and support Sun-centred astronomy, he was mistaken.

The story gets increasingly murky. Perhaps Barberini was doing his best to be both a cautious Church ruler and a friend to this brilliant, volatile man. Perhaps he still hoped to let matters evolve more gradually. Would a more politically sensitive and diplomatic person than Galileo have recognized that writing an impartial book was as good as it was going to get, very good indeed in view of the 1616 decision? Such a book might have achieved a significant victory for Copernicanism, without requiring him to become a martyr in the process.

Galileo was almost surely not thinking of martyrdom as a possible outcome. What *was* he thinking? He might quite understandably have decided, in view of the Church's earlier policy of tolerance, that the 1616 judgement had been only a temporary conservative aberration. Significantly, he now thought he had discovered the proof that he had been warned to find when that judgement was delivered – his new theory of the tides. It's possible he took Barberini's statement that as Pope he could no longer chat off the record to mean that he couldn't tell Galileo outright to author a pro-Copernican book but would like to see it happen. If Galileo *had* produced a less pointedly pro-Copernican book, would his fate or the future of Copernican theory *vis-à-vis* the Catholic Church have been different? Arguably not, for Galileo's trial as it turned out was a trumped-up charade in which science, religion and even Church authority were paid little more than lip service. What was really at the heart of the problem? That is still a mystery.

In any case, Galileo's book, though it was to seal the triumph of Copernican astronomy in the long run, in the short run played directly into the hands of his enemies. *Dialogue Concerning the Two Chief World Systems*, more often called by its shortened Italian name, *Dialogo*, makes a powerful case for Copernicus. Unlike *De revolutionibus*, it is not a technical,

mathematical book. It is an entertaining masterpiece of popu-
larization, written in Italian rather than in scholarly Latin,
designed to appeal to a great many readers – as no book on
Ptolemaic astronomy, indeed no book on astronomy, ever had
before. It takes the form of a lively four-day discussion among
three friends. The first day is used to demolish the ideas of
Aristotle. The second and third days are devoted to proving
that the Earth turns on its axis and orbits the Sun. The fourth
is spent on Galileo's theory of the tides, a theory which is in
error but which he thought was his clinching argument.
(Galileo in fact called Kepler childish for supposing that the
Moon affects the tides.)

Galileo left no doubt whatsoever that the character 'Salviati',
who is a far more intelligent man than either of the other
characters and who argues eloquently for Copernican theory,
represents Galileo himself. Salviati has the floor much of the
time. The character who argues the Aristotelian/Ptolemaic side
of the case – Simplicio – gets the last word, but he is little
better than a buffoon, confused and slow. One is tempted to say
that he provides the comic interest. Lest any of Galileo's readers
miss the point, a third character, Sagredo, a skilful discussion
leader who asks thoughtful questions to move the conversation
along, scolds Simplicio for his stupidity and inability to see
reason and congratulates him condescendingly whenever he
does see the light just a little. Galileo was a master of ridicule,
and he used that weapon mercilessly in *Dialogo*. Doubtless it
seemed to him that anyone who would continue to defend
Aristotle and Ptolemy in the face of compelling arguments for
Copernican astronomy *could* only be a half-wit. Galileo surely
could not possibly have thought his book was impartial or that
anyone else would think it was.

How amazing then to learn that there was initially no adverse
reaction! The book made its way through Church bureaucracy
with only a few annoying delays which only with hindsight look
ominous. After minor changes, it actually won approval from

the Church censors, who made one inadvertent improvement by requiring Galileo to change the title from *Dialogue Concerning the Tides* to *Dialogue Concerning the Two Chief World Systems*. The censors had no way of knowing that the section of the book about the tides was its one embarrassing weakness.

It might begin to seem at this point that Galileo had judged the situation entirely correctly. *Dialogo* appeared in February 1632 with a great flurry of publicity, and the reception was overwhelmingly enthusiastic. There was, to be sure, some criticism, notably from one Christoph Scheiner, a Jesuit who had earlier clashed with Galileo about which of them had first discovered and correctly interpreted Sun-spots. Galileo had pretty well trounced Scheiner in that encounter and was not gracious about the victory. It comes as no great surprise that Scheiner was more than a little sour about Galileo's latest triumph.

Then, out of the blue, disaster struck. Barberini, Pope Urban VIII, Galileo's old friend, suddenly turned bitterly against him. To this day Barberini's belated change of heart remains unexplained. Had he only just got around to reading the book? The argument that the Pope was under political and military pressures that caused him to change his mind about Galileo is unconvincing. These problems were so far removed from Galileo and the Copernican question that it is difficult to imagine they had an impact except to contribute to a state of tension. It also makes little sense to say that the Church could not be seen to waver on any decision. On what decision was it wavering? The 1616 dictum? Copernicanism had not been officially declared heresy then, and lest anyone had missed that technicality, much more recently Church censors had approved Galileo's book. Why waver from *that* decision? Barberini seems to have unleashed the fury of the Church against Galileo *in spite* of the fact that he must have known the result would be to make the Church and himself look vacillating and foolish. Many devout liberal Catholics were shocked when the Church finally condemned and prohibited the teaching of Copernicanism, because it committed the Church to a theory that now looked untenable. Barberini would have anticipated this reaction.

One popular explanation is that advisers around the Pope convinced him that Galileo had meant Simplicio to be a caricature of Barberini himself. Galileo had made the mistake of putting in Simplicio's mouth an argument that Barberini had once expressed in their discussions. Even so, the reaction seems extreme. The interpretation that makes this episode into a major confrontation between science and religion is that the Pope saw Galileo and Copernicanism as a threat to belief in the validity of scripture and to the Church's right to be the final arbiter of truth. In fact, such motivation might have pushed a good politician, which Barberini was, to side with the theory that looked likely to win the day. Approving Galileo's book was a strong move for the Church. A more likely reason for Barberini's ire is more subtle – that Galileo had finally managed to make it appear to the public that the Church had thrown its weight solidly behind Copernican astronomy. The decision to do that should have come, if it came at all, from Church officials, not from Galileo. This issue had little to do with the Church's right to declare what was true. It had to do with its right to choose its own political moment. In this respect, Galileo had indeed outmanoeuvred the cautious Church hierarchy and usurped Barberini's power.

When the printer of *Dialogo* received the surprising order to send all unsold copies of the book to Rome, he couldn't comply. They were sold out. But the news soon got about that *Dialogo* was to be re-examined to determine whether it was heretical. The Holy Office sent for Galileo. He went to Rome and resided at the home of the Tuscan ambassador for some weeks while the Holy Office scrambled to put together a coherent case – no easy matter, since Church censors had found no fault with Galileo's book and the opinion that Copernican theory was heresy had never been the official opinion of the Church. Church legal experts nevertheless did their best (or worst), and Galileo stood trial for 'a vehement suspicion of heresy'. Galileo's letter from Cardinal Bellarmine that said Galileo had not been forbidden by the 1616 decree to teach Copernicanism carried little weight, though it did cause chagrin. Even less effective was Galileo's pitiful defence strategy. He told the court that his accusers had misinterpreted *Dialogo*. The book actually favoured Ptolemy, not Copernicus, and he could add some pages at the end to make that clear. Even the least astute among the inquisitors could not be expected to believe this, for Galileo had, in *Dialogo*, called defenders of Ptolemy 'dumb idiots'.

Galileo was not tortured nor was he sentenced to death. It seemed to be his complete humiliation that the Pope wanted. He was forced to renounce Copernicanism in a long, demeaning statement before a resplendent crowd of dignitaries, nearly all of whom must have known the old man didn't mean a word of it. The story goes that he made his exit muttering '*eppur si muove*' ('and yet it does move'). Whether or not he really said the words out loud, he must surely have been thinking them.

Galileo was sentenced to spend the rest of his life in isolation, under house arrest at his own villa at Arcetri. Some scholars argue that the only thing that saved him from a harsher punishment was that Barberini, for all his inexplicable blind rage, *was* smart enough to recognize that this was going to be a Pyrrhic victory, that history and many of his contemporaries within Church officialdom would ridicule him for upholding Earth-centred astronomy. As it turns out, history remembers him as a cruel, irrational bigot. His Church and Catholic scholarship and science suffered irreparable damage and loss of credibility. After the trial, Italy lapsed into what was almost a scientific dark age with the prohibition of Copernican theory. The centre of scientific endeavour and achievement shifted to northern Europe and England, never to return.

Galileo lived for eight years after his trial. He was in his seventies, but still active. It was during these years that he got around to publishing much of the scientific work that he had carried out when he lived in Padua. He died in 1642, at the age of 78.

Did Galileo know that though he had suffered personal defeat he had won the war for Copernicanism? Probably. It's less probable that he realized he would go down in history as a scientific martyr and a symbolic figure for all those who see religion as the enemy of science, and vice versa. He would not have liked that, surely. For he believed so firmly and argued so well, himself, that there was no contradiction between them.

As Kepler, Tycho Brahe, Galileo and other 16th- and 17th-century scholars made up their minds whether to become Copernicans, some less-than-familiar scientific values came into play. It certainly wasn't always just a question of what fitted best with observation – not even for Galileo, who has been dubbed the father of modern science. Why else prefer one model to another?

Suppose that you and I, for reasons we won't pause to examine, have chosen the Moon as the centre of the system. We know that the issue here is one of relative motion only, and that we can come up with a Moon-centred model that does indeed fit with all the data and that no one can prove is wrong. What then *would* make anyone prefer another model to ours? What has Copernicus got that we haven't?

Imagine our planetary system represented as dots moving on a three-dimensional super-computer screen. Freeze first one dot and then another, each time allowing the computer to make sense of the movement of the other dots in terms of the unmoving dot. The picture may always be correct and accurate, but it will sometimes look far more complicated, and sometimes far simpler. The choice of one particular dot, the

Sun, as 'centre' causes the picture to fall into place and appear remarkably simple and harmonious. We feel as though we have cracked the code. Why bother with the other ways, just because they don't happen to be 'wrong'? This reasoning causes scientists to prefer the Copernican model to Tycho Brahe's.

Simplicity and harmony are strong pointers but not absolute clinchers of the sort Mr Elmendorf was hoping for. There are other criteria. Modern science obliges us to back up a choice of *how* the solar system moves with the answer to a second question: *Why?* It isn't enough to claim that a model can predict where all the planets will be a month or a year or a century from now. What makes them go there? To use more scientific language, what are the 'dynamics' that cause them to move in this way rather than another? What makes them move at all rather than sit still? Ptolemaic astronomy explained the movement of the universe as originating in the sphere of the stars, with that movement transferred to the planetary spheres. There were those who thought that the Creator had given an initial push to each planet at the moment of creation, and that that movement would go on forever. There were others who suggested that angels move the spheres. Copernicus didn't attempt to explain why the planets moved as he thought they did, although he recognized the importance of the question. Kepler believed that a whirling force emanating from the Sun drives the planets and that this could work only if the Sun was at the centre. But not until Isaac Newton and later Albert Einstein, with their explanations of gravity and the curvature of space time, was anyone able to suggest the reasons accepted today why the planets move as they do.

Is that how we know Copernicus was right? Not quite. Fred Hoyle has argued in his book about Copernicus that a subtler understanding of Einstein's theories reveals that they may actually slightly favour an Earth-centred model. Had Galileo had Hoyle at his elbow, he might have produced the book that would have pleased the Pope and not have been tried for heresy!

Why, then, should poor Ptolemy lose out so badly? Paradoxically, the enormous success of Ptolemaic astronomy is not an argument in its favour. It can account for all apparent movement in the heavens. It could also account for a great deal that we never see happening. It allows for too much. Copernican astronomy, as it has evolved, allows for far less. It is easier to think of something that Copernican theory could not incorporate. The more scientific way of putting this is that Copernican theory is more easily 'falsifiable' than Ptolemy's, easier to *dis*prove. Falsifiability is considered a strength. If a theory sets up a clear enough profile so that it provides numerous opportunities to shoot it down, and no one is able to shoot it down – if new discoveries don't undermine it but fall neatly into place, as Galileo's discoveries with his telescope did – if the picture gets more harmonious over time rather than more complicated, then a theory begins to look as though it is right on target.

There is another criterion by which theories are judged. For better or for worse, it shows that modern scientists do have a certain kinship with those recalcitrant 17th-century scholars they so disdain. When new theories and the implications of new discoveries disagree with the way a scientist personally feels the universe ought to run, he or she is reluctant to accept them. In the 20th century, Einstein's resistance to the notion of an expanding universe, as well as many scientists' discomfort with the possibility that the universe may be far less predictable than previously thought, are examples. One need not be narrow-minded or anti-science to dismiss a theory on the grounds that it offends one's scientific aesthetic sensibility. 'It can't be right, because it feels so wrong.'

These few paragraphs have viewed the competition with the eyes of 20th-century philosophy of science, at the end of a century that has given an extraordinary amount of thought to the question of how we know what we know. But modern thought processes didn't spring out of nowhere. They are rooted in thinking that was already substantially in place even in the ancient world and certainly in Kepler's and Galileo's time. Scholars didn't think in terms of relative motion in the 20th-century sense, but they knew that there could be competing and equally valid geometric explanations. Those who opposed Galileo used this line of argument. When it came to making simplicity a criterion, scholars had no 'scientific method' yet to tell them that the most economical and harmonious model that saves the appearances is the best model. But even the ancient astronomer Hipparchus was known as a man who always preferred the least complicated hypothesis compatible with observation, and Neoplatonism brought into late-medieval science a preference for finding simple mathematical and geometric regularities in nature. Copernicus firmly believed in the superior harmony of his arrangement and saw that as a powerful argument in its favour. The search for such harmony was the driving force behind Kepler's work. Some scholars expressed this preference in religious terms: the work of God and nature is work of great elegance and simplicity, and we must try to understand and explain similarly.

Modern scholars can never know what really happened in the case of Galileo or unravel all the currents and cross-currents of thought (and lack thereof) and behaviour that were part of that chain of events. The surface arguments hid ambiguities, assumptions and personal agendas that would have been impossible to unmask even had we been there in person. Nor can scientists and philosophers of science tell precisely what causes one theory to emerge as 'scientific knowledge' while another is consigned to the dustbin. The stories of Ptolemy, Copernicus, Kepler, Tycho Brahe and Galileo should at least discourage simplistic interpretations in terms of proved/unproved, bad science/good science, knowledge/ignorance, pro-Ptolemy/pro-Copernicus, religion/science, open-mindedness/blind dogmatism – and caution us not to trust so readily the popular legends that have grown up around historical figures. Indeed, trying to unravel the complex human saga of the Copernican revolution makes explaining the night sky seem a relatively simple, straightforward task.

CHAPTER 4

An Orbit with a View
1630–1900

What a wonderful and amazing scheme have we here of the magnificent vastness of the universe!

Christiaan Huygens

When French astronomer Pierre Gassendi watched the transit of Mercury in 1631, he couldn't believe what he saw. Gassendi, along with the rest of his generation, had studied the cosmic dimensions given by Ptolemy. The tiny dot on the Sun was surely a sunspot, not the planet. Kepler too had expected a much larger apparent diameter for Mercury. Only when Gassendi realized that the dot was moving much too quickly across the face of the Sun to be a sunspot did he conclude with astonishment that, in spite of its 'entirely paradoxical smallness', this was indeed Mercury. In fairness to Ptolemy, he had commented in his *Almagest* that the absence of observed transits of Venus and Mercury might be the result of their apparent smallness. Part of the reason Ptolemy often comes off so badly is that astronomers didn't read him very carefully!

Ptolemy thought of his cosmic dimension estimates as a

speculative offshoot of his mathematical astronomy, but it's evident from both scholarly writing and popular literature that the Middle Ages regarded them as established truth, and Ptolemy's numbers still had a tenacious hold on the minds of scholars in the early 17th century. However, in the 1630s, when astronomers like Gassendi began using telescopes on a day-to-day basis, and with the dissemination of Kepler's laws, the ancient distance measurements finally lost credibility, at first among specialists and then among all educated people. Ptolemy's distances were far too small.

Mercury's transit in 1631 permitted the first accurate measurement of its apparent diameter (that is, how large its diameter appears from Earth), and eight years later there was a similar opportunity to measure Venus.

The quality of telescopes was improving, and there were also advances in related technology. In the spring of 1610, Galileo had been using a telescope that magnified thirty times. Two years later, astronomer Thomas Harriot had one that magnified fifty times, but at this magnification the area visible through it was frustratingly small. Then in the 1630s, an instrument known as the astronomical telescope came into use. Kepler had first introduced the theory of this telescope in 1611, but his version inverted the image. When changes in the configuration of lenses solved this problem, the astronomical telescope came into its own, providing a much larger field of view at high magnifications.

The astronomical telescope had another great advantage. Yorkshireman William Gascoigne discovered it by accident in the late 1630s when a spider spun a web in his telescope and he saw some of the threads of the web sharply outlined against the background image. What this told Gascoigne was that an object inside an astronomical telescope could appear in sharp focus, superimposed on a distant object under observation. In other words, he could place some sort of ruler inside his telescope. The ruler he built was a 'micrometer' (as the name suggests, an instrument for measuring extremely small dimensions), with cross-wires that moved across the image with the turning of a screw. Before this invention it was possible only to estimate the apparent size of a body seen through the lens, and these estimates varied widely. Now there was a way to *measure* the apparent size, comparing it directly with an established stand-ard. Sadly, Gascoigne died in the battle of Marston Moor in 1644, and his invention remained unknown to other astrono-mers until the 1660s. By then French experts had developed a similar idea from astronomer Christiaan Huygens. The Royal Society in England were quick to claim Gascoigne's prior discovery, but Frenchmen Adrien Auzout and Jean Picard were largely responsible for developing the screw micrometer into a fine precision instrument.

By 1675 observers were beginning to agree on the apparent sizes of the planets. However, consensus about their distances, now that Ptolemy was no longer to be trusted, was slower in coming. Kepler's third law of planetary motion had established a relationship among the planets having to do with their orbital periods and their distances from the Sun, but Earth-bound observers were still little better off in terms of absolute measurements than if they had been standing at the foot of a ladder leading up into the sky, knowing that the second rung was twice as far as the first, and the third three times as far, and so forth, but having no clue how far up the first rung was.

In the southern wall of Bologna's Cathedral of San Petronio, there is a small opening high up through which a shaft of sunlight penetrates to the cathedral floor below. There, a gnomon – a protrusion like the raised part of a sundial – makes possible the measurement of the shifting image of the Sun. In the mid-17th century, alterations to the cathedral made the old gnomon useless, and Gian Domenico Cassini of the University of Bologna was given the task of setting up a new one to replace it. Cassini was a good choice for the job, for he was keenly interested in the Sun and its movements. Cassini's gnomon still stands in the cathedral.

In 1669, when Cassini was 44, Jean Baptiste Colbert, an influential minister in the court of King Louis XIV of France and member of the recently founded Academy of Sciences in Paris, invited him to come to Paris to work at the new Royal Observatory. Louis XIV was the king who built Versailles, engaging another eminent Italian, architect and sculptor Giovanni Bernini, for that task. Louis cultivated a public image of himself as the Sun King and surrounded his sumptuous court with appropriate symbolism. Naturally, his chief astronomer should be not only one of the best in Europe but also an expert on the Sun.

There were additional, scientific, reasons for bringing Cassini to Paris. The Academy were already deeply interested in solar theory and eager to gain a better understanding of the Sun, the planets and the 'refraction' of light by the Earth's atmosphere – that is, the way the atmosphere bends and smears light rays passing through it. It was their intention to bring astronomical tables into line with new measurements provided by telescopes and micrometers. Cassini accepted the invitation and became head of the Observatory. The French version of his name was Jean-Dominique Cassini.

All over Europe at this time telescopes were getting longer and longer. Some experts even voiced the extravagant hope that a long enough telescope would allow astronomers to study the animals on the Moon. One of Cassini's first undertakings was to make sure that the new Observatory acquired some of the finest instruments available. In 1671, he brought from Rome a 17-foot telescope manufactured by his good friend and former colleague Giuseppe Campani, one of the most skilled telescope builders in Europe. Colbert, the courtier at whose invitation Cassini had come to Paris, was particularly taken with this telescope, and his enthusiasm prompted the King to present the Observatory with a still larger Campani telescope, a 34-footer, and later, in 1684, a 100-footer. The old wooden Marly water-tower was trundled over to the Observatory to serve as a support for these giant instruments, with steps built to the top and a rail added to prevent astronomers and their assistants from toppling off on dark nights. Among the discoveries Cassini made with these telescopes were several moons around Saturn and the dark division in Saturn's rings. His most famous achievement, however, was the first successful measurement of distances within the solar system.

Cassini and other astronomers knew that in August and September of 1672, Mars would be at its nearest proximity to the Earth, and this would provide optimum conditions for measuring its distance. The method Cassini planned to use was an old mapmaker's trick, known as triangulation, in which two observers in two widely separated locations measure the position, against the background, of the same distant object at the same time. Because they are far apart, the two observers have different viewing angles.

Human beings and other animals use this technique instinctively to judge distances. Our two eyes are the two 'observers'. The view from the right eye is not exactly the same as the view from the left. The demonstration is familiar: hold a finger upright close in front of you. Shut one eye, then the other, and the finger appears to change position relative to the background, although it actually hasn't moved. The closer your finger is to your face, the greater the shift will appear to be. It doesn't occur to us to use geometry to measure the distance from our finger to our face from how great the apparent shift is. Our brains do it automatically with a precision that is adequate for most everyday purposes.

The apparent shift in the position of an object against the background, as seen from two different locations, is called a parallax shift. To understand the use of this shift for measuring distances: suppose you are driving along a road across a desert. The desert stretches in all directions to the horizon, where a string of mountains is just barely visible. There is a solitary cactus not far from the road. The problem at hand is how to measure the distance from the road to the cactus without leaving the road.

Begin by constructing two imaginary triangles. The first triangle has the road as one of its three sides and a line drawn straight to the cactus as a second side. See Figure 4.1a. The letter X designates the point at which the line to the cactus meets the road. Standing at X, find a landmark on the horizon that is directly behind the cactus. Fortunately, there is one, a snowcapped peak that stands out clearly from the others. From X, move a little to the left along the road. Look again at the cactus and the mountain peak. They will no longer appear to be lined up. The peak will have moved out from behind the cactus. Call the location where you have stopped to take a second look 'Y'. (To associate this analogy with the finger exercise above: X is like the view from your right eye, and Y is like the view from your left.)

Standing at Y, the next step is to build the second imaginary triangle. One corner of this triangle is the point where you now stand (Y). One side of it is an imaginary line drawn from you at Y *through* the cactus off to the horizon. There is a convenient landmark there as well, a particularly precipitous crag. You don't have to know the length of the line to the crag. A second side of the triangle is an imaginary line drawn from you at Y through the peak. Again, you don't have to know the length of the line. See Figure 4.1b. What you do need to know is the 'angular distance' between the peak and the crag.

Angular distance is the number of 'degrees' between two lines pointing from an observer (you at Y) to two distant objects whose separation the observer wishes to know (the peak and the crag). It works like this: if you are at the centre of a circle, and the crag and the peak are on the rim of the circle, how many degrees apart on that rim are they? A circle has 360°. A quarter way around a circle (from 12 o'clock to 3 o'clock) is

Figure 4.1

a. The cactus is near the road and there is a ring of mountains on the horizon. From X, on the road, the cactus is directly lined up with a snowcapped peak. Y is a little to the left of X. There is an imaginary triangle whose sides are a line from X to Y, a line from X to the cactus, and a line from Y to the cactus.

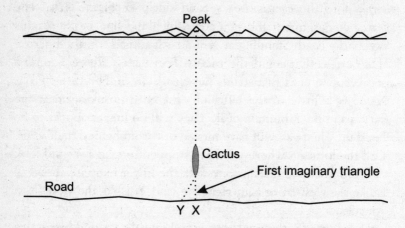

b. Draw lines from Y to the peak and from Y through the cactus to the horizon. The second line is a continuation of the line previously drawn from Y to the cactus, and it leads to a crag on the horizon.

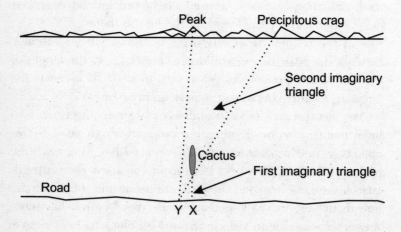

90°. (Lines drawn from 12 and 3 o'clock to the centre of the circle – where the observer is supposed to be – meet there at a 90° angle.) The angular distance between the peak and the crag is obviously not that great. It's more like 12 o'clock to 1 o'clock. That's 30°.

Now pretend you are hovering in a helicopter over this scene, for it's necessary to draw an idealized picture of it from that vantage point. On Figure 4.2, the large circle is the horizon, with the mountains. The dot in the centre of the circle is the cactus. The segment of the road between X and Y is part of a smaller circle, centred on the cactus and near to it. We don't yet know the size of that circle, but that doesn't matter at the moment. We've drawn it in as a broken line. In Figure 4.2, the huge circle of the horizon becomes a giant clock face, aligned so that the snowcapped peak is at twelve o'clock. The two lines that meet at Y, on the road, end up widely separated at the horizon. One line, of course, still leads to the peak, the other leads to 'one o'clock', where the precipitous crag is. The clock face gives us a way to measure the separation between the two landmarks, and, hence, the angle of separation between the two lines. Twelve o'clock to one o'clock on a clock face (viewed from the centre, where the hands are attached) is a shift of 30°, meaning that the angle at Y (close to the centre), where those two lines meet, is approximately a 30° angle.

So far you may not seem to have found out very much. Actually you have all the information necessary to calculate the distance to the cactus from the road.

Taking stock: you can pace off the distance between X and Y on the road – an easily measurable distance. You know the angular separation between the two lines leading from Y to the peak and to the crag on the horizon (30°). The rules of geometry say that you, standing at Y, will measure almost exactly the same angular separation between the peak and the crag as the cactus, looking back at the road, would measure between X and Y. On the clock-face drawing, the lines drawn

109

Figure 4.2

An idealized drawing, from the air, shows the cactus at the centre of a huge circle (the horizon), visualized as a clock. The part of the road between X and Y is a segment of a much smaller imaginary circle centred on the cactus. Lines drawn from Y to the peak and through the cactus to the horizon form approximately a 30° angle at Y.

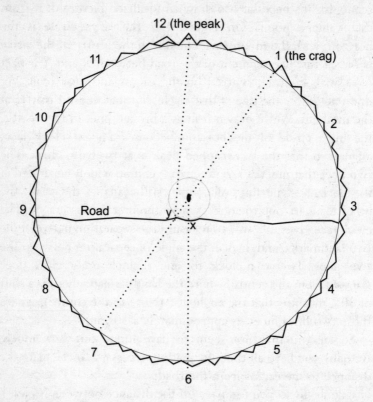

from the cactus through X and Y pass approximately through six o'clock and seven o'clock, and that means that the angular separation between those two lines, also, is approximately 30°. If the horizon were as far away from the cactus, X, and Y as the stars are from Mars and the Earth, the angles would be even closer to identical.

Continuing, then, with the measurement: there is only one distance the cactus *could* be from the road at which the cactus would measure a 30° angular separation between X and Y, and at which the *actual* distance between X and Y along the road would be the distance you paced off.

To apply the same method to the distance to Mars: the cactus is Mars, and the horizon with the ring of mountains is the distant stars. It's necessary to find two viewing positions analogous to X and Y. Astronomers must be able to get to both positions to take measurements and they must know the distance between the viewing positions. The two positions must be far enough apart so that a parallax shift can be detected. The distance between them – called the 'base line' for the measurement – is a tiny part of an imaginary circle analogous to the little one in Figure 4.2. In the Mars measurement, this imaginary circle is so minuscule compared to the ring of distant stars that Mars, X and Y can be thought of as all being, essentially, at the centre of the star ring. The parallax shift of Mars as viewed from two ends of any base line possible on Earth is not anywhere near as large as 30°.

It should be clear now why Cassini's project required having observers in two widely separated locations on the face of the Earth. There had been vague plans for the Observatory to send an expedition to the tropics for other astronomical research, and this rare opportunity to make the first-ever measurement to another planet was an incentive to proceed posthaste with those plans. Fortuitously, Colbert was hoping to establish a colony at the mouth of the Cayenne River in South America in what is now French Guiana, where there had been French settlers earlier in the century. Ships were sailing there regularly. Cassini sent a young colleague, Jean Richer, and an assistant, a M Meurisse, off to Cayenne equipped with several measuring devices and telescopic sights. They were instructed by the Academy to make observations leading to the measurement of the parallaxes of the Moon, the Sun, Venus, and especially of Mars.

In Cassini's triangle (analogous to the 'first imaginary triangle' in Figures 4.1a and 4.1b) the three corners were Paris, Cayenne and Mars. He knew approximately the distance from Paris to Cayenne (the longitude of Cayenne was not certain, but that was something the expedition hoped to remedy), taking into account the curvature of the Earth. That distance was the base line, Cassini's 'distance from X to Y'. It was possible, though problematical with the instruments available to Cassini, to detect the parallax of Mars from that base line. Parallax measurement can, in principle, be made to work for anything that can be seen to have a parallax shift. Figure 4.3 demonstrates (though not to scale or with the angles that really exist) parallax measurements of the Moon and Mars.

Cassini's measurement to Mars was not as simple as finding the distance to a cactus, for several reasons. A cactus isn't going anywhere. A planet is. If Cassini and Richer couldn't time their measurements precisely and know how the time of a measurement in Cayenne compared with the time of a measurement in Paris, the resulting data would be worthless. Cassini couldn't merely give the order 'synchronize watches'. The best clocks available were pendulums. They could be synchronized but they wouldn't stay synchronized while one of them was taking a sea voyage to the other side of the world.

During his years at the University of Bologna, Cassini had studied the eclipses of Jupiter's moons and their shadows as they crossed the body of the planet, and he had drawn up tables of their motions. In 1666 he'd noticed that the moons were close enough to Jupiter so that a moon's appearance from behind Jupiter would be seen simultaneously from any point on Earth, and he realized that Jupiter and its moons could provide a way to determine the time difference between widely separated points on the Earth.

Other problems couldn't be so handily solved and caused some later experts to be critical of Cassini's blunt announcement of a definite parallax for Mars, for Cassini was well aware of the

Figure 4.3

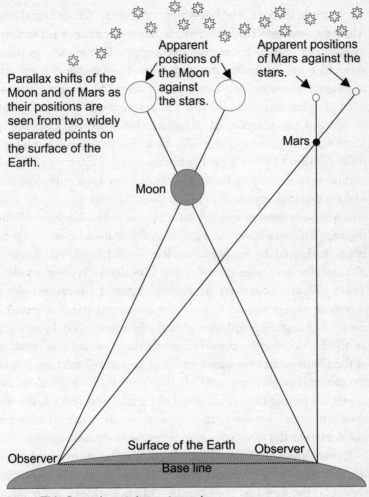

Parallax shifts of the Moon and of Mars as their positions are seen from two widely separated points on the surface of the Earth.

Apparent positions of the Moon against the stars.

Apparent positions of Mars against the stars.

Mars

Moon

Surface of the Earth

Observer   Observer

Base line

Note: This figure is not drawn to scale.

inevitability of a large margin of error in his measurements. In fact, probably no one was better able to appreciate that margin of error than he, because of his earlier study of the refraction of light

113

by the Earth's atmosphere and his previous efforts to measure the Sun's parallax.

Nevertheless, in the summer of 1673, Cassini and the Academy awaited Richer's return from Cayenne with intense excitement. When he arrived, in August, Cassini set immediately to work analysing the data from Cayenne and Paris and additional locations in France where there had been observations. He found most of it not too revealing, but when he had completed his analysis, he reported that from the parallax of Mars he had derived a distance from Earth to the Sun of 87 million miles (140 million kilometres).

There were others besides Cassini who took advantage of Mars's unusual proximity. In England, a young, largely self-educated astronomer named John Flamsteed had been studying the Sun, Moon and planets. He decided that an old method Tycho Brahe had used to measure parallax would work much more successfully now that there were telescopes. Tycho's method didn't require observers at widely separated locations. One observer would suffice to measure a 'diurnal parallax', which means a change in position as observed from one single location on the Earth's surface at two different times of day. The rotation of the Earth carries the observer from one end of the base line to the other. Though Flamsteed's father sent him on a business trip at just the wrong time, Flamsteed did manage to observe on one clear night and he also came up with parallax calculations for Mars. His results were largely in agreement with Cassini's.

Cassini's and Flamsteed's findings did not, as is often supposed, bring immediate consensus among astronomers about the distances to the Sun and the planets. Others besides Cassini were aware of the large margin of error. However, though these measurements were still imprecise and somewhat in dispute, they became the key to the solar system. Now it was possible to use Kepler's laws to calculate distances from the Sun to all the known planets. It was a source of amazement how far away the Sun was from Earth. Copernicus had calculated the distance from Earth to the Sun as two million miles; Tycho Brahe, five million; Kepler, no more than 14 million. The new measurement was 87 million miles or 140 million kilometres. (The modern measurement is 93 million miles or 149.5 million kilometres.) In the late 1670s, men and women thus for the first time became aware of the size of the solar system and found it enormous beyond anyone's previous imagining. The universe beyond must be inconceivably vast. In the Middle Ages the distance to the 'fixed stars' had been illustrated by how many years Adam, starting his journey on the day of Creation, would still have to walk at a rate of 25 miles per day to reach them. We hope he brought provisions for a long hike. The new illustration had a cannon ball travelling at 600 feet a second (considerably faster than Adam, and presumably passing him as it went) taking 692,000 years.

Cassini went on attempting to measure parallaxes and to tease fresh results out of the data from the Cayenne expedition. His work continued to fascinate the King, the court (especially the faithful Colbert) and the public. He became, for a while, *the* dominant figure in astronomy in Europe. Like Galileo, Cassini was astute when it came to self-promotion. More than anyone else in his time, he used the publication of papers in scientific periodicals as a way of establishing priority and of letting the educated world hear of his successes.

In 1676, Jupiter's moons made possible another measurement that would be essential to later astronomy. Ole Roemer, a Danish astronomer also working at the Royal Observatory in Paris, studied the eclipses of Jupiter's moons and noticed that the time that elapsed between their disappearances behind Jupiter varied with the distance between Jupiter and the Earth, a distance that changes as the two planets move in their orbits. Roemer speculated that the velocity of light was responsible for what looked like delays in the eclipses. When Jupiter was further from the Earth, it took longer for the picture of the eclipse to reach eyes and telescopes on the Earth. Timing the delays, Roemer proceeded to calculate the speed of light at about 140,000 miles per second. That figure fell short of the 186,282 miles per second that we assign to the speed of light today. The discrepancy in the measurement was due in part to Roemer's less-than-precise knowledge of the distance to Jupiter. The modern calculation is made with an atomic clock and a laser beam. (This book uses rounded-off figures of 186,000 miles or 300,000 kilometres per second for the speed of light.)

The Royal Observatory in England was established somewhat later than the Paris Observatory, and the impetus for its founding came from an unlikely source. In 1674, Louise de Kéroualle – a Bretonne who had recently been made Duchess of Portsmouth and was one of King Charles II's mistresses – brought a Frenchman named Sieur de St Pierre to the King's attention. St Pierre had discovered, so he said, a secret method for finding longitudes. Determining longitude requires having some form of universal clock – such as Jupiter's moons were for Cassini and Richer – that will allow comparison of celestial phenomena as seen from different locations. If the Sun is directly overhead in New York, what is its position in Greenwich, England? Answering such questions is one way of finding longitude.

Flamsteed and others thought that St Pierre's 'secret method' must involve using the Moon as a time-keeper. Who needed St Pierre? King Charles told the Royal Society, then known as the 'Royal Society of London for Improving Natural Knowledge', to collect whatever lunar data were necessary to determine whether the Moon could serve as a universal clock. Flamsteed soon reported that lunar and stellar positions were not known sufficiently well to make the method reliable. Nevertheless, the bee was in the King's bonnet and he founded the Observatory, designating Flamsteed as 'astronomical observator', a position that would later become 'astronomer royal'. A new building, designed by Sir Christopher Wren, was ready for occupancy in July 1676, and Flamsteed and his associates began the task of producing correct star positions and tables for the Sun, Moon and planets – for the express purpose, as those who governed England saw it, of aiding navigation, but also for the furthering of astronomy.

So things stood in the last quarter of the 17th century. Kings were footing the bill for observatories in Paris and at Greenwich. There was also royal patronage for such bodies as the French Academy and the Royal Society, where gentlemen – expert and not – got together and shared knowledge of the wonders of science. They had a fairly good idea of the dimensions of the solar system. Astronomy seemed poised and ready to measure much greater distances. But a base line from Paris to Cayenne isn't long enough to measure the parallax shift of stars; in fact, no possible base line on the face of the Earth would suffice. Was there any way to make the parallax method work for stars?

The clue was there long before Ptolemy. One objection other Hellenistic scholars raised to Aristarchus's suggestion that the Earth moves and orbits the Sun was that, if it does, we ought to be able to observe stellar parallax as we travel from one extreme of the orbit to the other. In other words, the stars ought to shift position relative to one another in the sky, just as the cactus and the mountain peak shifted. If we're viewing the cactus and the mountains from a car window and they don't seem to be changing positions relative to one another, either our car is standing still or both the cactus and the mountains are very far away. Likewise either the Earth does not move and orbit the Sun, or the stars are extremely far away – further away than most ancient scholars, except for Aristarchus, were willing to conceive.

By 1700, nearly all astronomers agreed that the Earth rotates on its axis and orbits the Sun. They also recognized the potential of this orbit for a base line. Even so, at that time and even much later, no one was able to detect an annual shift in a star's position. That discovery had to wait for considerable improvement in telescopes and the precision instruments used with them – advances that would eventually come with the Industrial Revolution. But there was also meanwhile more to be learned from a theoretical point of view, in order to be able to recognize stellar parallax and not confuse it with other effects.

Parallax wasn't the only hope for measuring the distance to the stars. According to the 'inverse square law', the brightness of a light falls off with distance in a mathematically dependable way. The measured intensity of light diminishes by the square of the distance to its source. Imagine that you have two 100-watt lightbulbs. Place one of them twice as far away as the other. The further bulb will appear to be only a fourth as bright as the nearer. It will seem to you that it must be a 25-watt bulb. Turning that exercise around: suppose you keep one of the 100-watt bulbs nearby and ask an assistant to carry the second to an unknown distance. If the more distant bulb appears to be a 25-watt bulb, then you can be sure it's twice as far away as the first bulb. Comparing their *apparent* brightness gives the distance of the second light. How does this apply to stars? If stars all have the same close-up brightness and we know the distance to one star, then in principle we can find the distance to any other star by comparing its *apparent* brightness (how it looks to us from the Earth) with the brightness of the first star.

The reasons why measuring the distance of stars has turned out to be far more complicated than that become clearer if we come in out of the dark and note that the same inverse square law that works with the brightness of light works with the size of objects. Even without knowing the mathematics, you and I use the method instinctively, if imprecisely. Suppose you are overlooking a vast expanse of land with a great number of elephants grazing on it. Some of the elephants look tiny, but because you have previous experience of elephants and know how large an elephant would be if you were standing near it, you make an educated guess that these aren't really tiny at all, just far away. From the difference between their apparent sizes and the normal close-up size of an elephant, you judge how far away.

Your success relies upon: (1) having some way of knowing that elephants are all approximately the same size – some reason for assuming there aren't elephants that are pygmies and others that are giants; and (2) having experience of what that size actually is – that is, how big an elephant looks when it's standing a known distance away.

With a light seen at night at an unknown distance, on the Earth or in the sky, we lack those essential ingredients. Experience tells us that lights can't be depended on all to have approximately the same close-up brightness, so knowing the close-up brightness of one light is of no use when studying the brightness of a distant light. A comparison is meaningless. We're left with such questions as: Is it a dim bicycle headlamp over there, just a few yards up the road, or is it the high-beam headlight of a car a mile away? Is it a meteor suffering a fiery death in the upper atmosphere, or is it a firefly no further away than the treetops?

The situation is puzzling but not completely hopeless, nor did it seem so at the end of the 17th century. Astronomers did know how far it is to one star – the Sun. Stars *might* all be equally bright, and it *might* be safe to assume the Sun is a typical star. It would seem highly desirable to know the brightness or the distance of at least one star other than the Sun, but failing that, the Sun would have to serve as a standard. On the assumption that it could, Englishman Isaac Newton set out to measure the distances to the nearest stars.

Newton, born in 1642, the year that Galileo died, was 30 years old when Cassini and Flamsteed measured the distance to Mars. Except for his *Principia Mathematica*, which appeared in 1687, unarguably one of the most important achievements in the history of human knowledge, Newton published almost nothing. However, he obsessively researched an astounding variety of subjects including optics, theology, alchemy and calculus, and was responsible for significant advances in many of the fields he investigated. He contributed to the development of a new type of telescope that used a mirror rather than a lens to focus incoming light. (See Figure 4.5 on page 134.)

Newton probably wouldn't even have published the *Principia* had his friend Edmund Halley not goaded him sufficiently. Publication meant public notice, and contact and correspondence with other scholars. It meant he would be expected to take part in the discussions at the Royal Society. There would be invitations to perform experiments for that body and to watch others do the same. Though honours were tempting and sometimes Newton succumbed, those distractions more often seemed anathema to him. They took precious time away from his research. It is surprising therefore that he agreed to become head of the Royal Society. Unfortunately, he used the position of power in an extremely unpleasant, autocratic manner, bringing misery to other fine scientists (including the elderly Flamsteed). Newton also eventually accepted the job as Master of the Mint and thoroughly enjoyed his rather pedestrian duties there.

Whatever his personal failings, it was Newton who capped the Copernican revolution with his discovery of laws of gravity. In Newton's description, each body in the universe is attracted towards every other body by the force called gravity. How much bodies are influenced by one another's gravitational attraction depends on how massive the bodies are and how near they are to one another. For instance, any change in the mass of the Earth or the Moon, or in their distance from one another, would change the strength of the gravitational attraction between them. If the mass of the Earth were doubled, the attraction between the Earth and the Moon would double. If the Moon were twice as far from the Earth as it is, the attraction of gravity between the Earth and the Moon would be only a quarter as strong.

Here at last in this simple description were the dynamics that cause the planets to move as they do rather than in some other way, the physical reasons behind Kepler's laws. The answer that eluded Ptolemy, Copernicus, Galileo and Kepler – but to which Kepler's laws point – was summed up in one sentence: the gravitational force between any two bodies is proportional to the product of their masses and inversely proportional to the square of the distance between them. It took Newton's genius to see that it is this same force – gravity – that keeps us from flying off the ground, dictates the path of a ball thrown on the Earth, underlies the motions of the planets, and governed the way Galileo's two objects landed at the foot of the Tower of Pisa (if they did).

Newton's 'laws of motion' could be tested by experiments as well as by astronomical observations, and this was an era that rejoiced in such testing. The scientific rigour that had made Galileo exceptional was beginning to be considered essential for any scientific activity. It became clear that Newton's formulae did indeed describe the way things happen. Nature was law-abiding and she was following these laws! It's difficult for us, who take for granted that science is able to predict reality and that simple, dependable mathematical and scientific rules underlie the apparent complications of nature, to appreciate how awe-inspiring it must have been for the many men and women who were realizing this in the 17th century for the first time. Not only did such laws exist, but human minds could discover them and understand them. The concept was not a complete novelty to scientists, although to see it demonstrated as beautifully as it was in Newton's *Principia was* a novelty. To the non-scientific public Newton's revelation was sensational, miraculous. The fame of his book spread quickly throughout Europe, and his ideas were popularized in many forms. There was a publication called *Newton for Ladies* in France. *Principia* did encounter some hostility on philosophical grounds, uneasiness with the notion that gravitation could act through empty space. 'Action at a distance' suggested the occult.

Newton's attempt to estimate the distances to the nearest stars by assuming that all stars, including the Sun, have approximately the same brightness, is one of his less celebrated efforts. In the analogy with the elephants, you and I assumed that all elephants are about the same size, but suppose we had seen only one elephant close up. It would be risky to jump to the conclusion that all elephants are the same size – that our local elephant is a typical elephant – based on such limited experience. The difference in apparent size between the local elephant and other elephants whose distance we don't know might actually be due to differences in size, not an indication of their distance at all. Newton's situation was even more ambiguous than that. Elephants look pretty much alike, but the Sun, as seen from Earth, doesn't resemble the other stars. Most experts in Newton's time did think that the Sun was a star. But was it a *typical* star? It wasn't unreasonable to decide to assume that it was, and see where that would lead.

Newton did that in an ingenious but rather convoluted way, using a technique suggested by Scottish mathematician and astronomer, James Gregory. Judging by the size of Saturn, Newton estimated that about one part in a billion of the Sun's light hits the planet Saturn. He reasoned that Saturn doesn't reflect all of the sunlight that hits it – so it would be incorrect to conclude that the light we see coming from Saturn represents one billionth of the Sun's light. Instead, he thought that Saturn probably reflects only about a quarter of the Sun's light that hits it, which would mean that the reflected sunlight coming from Saturn gives us a good indication of what one part in *four* billion of the Sun's light looks like. Following this line of thought, if a distant star seems to have the same brightness as Saturn, it follows that the light we are receiving from that star (not reflected sunlight; the star's own light) is also equivalent to one part in four billion of the Sun's light. What this means – *if* all stars' brightnesses are the same as the Sun's – is that a star that looks (from Earth) as bright as Saturn would have to be approximately a hundred thousand times further away from us than the Sun is.

Newton's measurements of the distances to some of the nearest stars were not far wide of the mark, though this method was not dependable partly because stars do differ in their absolute magnitude (their 'close-up' brightness), and a star's apparent magnitude (how bright it appears from Earth) alone can't be used as a gauge of its distance. (See box below.) Some of the stars that look brightest in the night sky are very far away, while many nearer stars are rather inconspicuous.

In 1718, Newton's younger friend Edmund Halley discovered an important new clue to star distances. He had become fascinated with Ptolemy's writings and the star catalogues that ancient astronomer had compiled. Halley was particularly curi-

---

The brightness with which a star would appear if it were only ten parsecs distant from us – what we might call its 'close-up brightness' – is its

absolute magnitude

How bright a star appears from the Earth as we view it in the night sky is its

apparent magnitude

It is the apparent magnitude of a star that falls off with distance. The absolute magnitude of a star does not change with distance.

A **parsec** is a little more than 30 trillion kilometres, or 3.26 light years.

A **light year** is the distance that light travels in one year – about 5.88 trillion miles or 9.4607 trillion kilometres.

---

ous as to whether the stars had changed positions since the time of Hipparchus and Ptolemy. He took it upon himself to compare positions recorded in Ptolemy's *Almagest* with positions in his own lifetime in the late 17th and early 18th century.

Halley was born in 1656 and while still an undergraduate at Oxford wrote and published a book on Kepler's laws. His book came to the attention of Flamsteed, who had a great deal of influence as the first Astronomer Royal of England (though the position wasn't called that yet) and first head of the Royal Observatory at Greenwich. Halley left Oxford without getting his degree and, at Flamsteed's behest, was soon on the island of St Helena in the South Atlantic, mapping the sky as seen from the southern hemisphere. When he returned to England two years later he was elected to the Royal Society. He was only 22.

Even more eclectic in his interests than Newton, Halley spent the next 30 years in an astounding variety of pursuits. He travelled extensively to meet other scientists and astronomers; he assisted Flamsteed; he got married; he commanded a warship in the Royal Navy and captained a mutinous ship across the Atlantic; he went to Vienna on two secret diplomatic missions; he served as deputy to the controller of the Mint at Chester (a position Newton helped secure for him); he studied magnetism and the winds and tides; he prevailed upon Newton to publish the *Principia,* and he financed its publication. Of course the work which brought him most fame was his study of comets. 'Halley's Comet' was named after him when it reappeared in 1758, after his death, at the time he had predicted. In 1703, Halley joined the faculty of the University of Oxford, where he had failed to complete his degree, as Galileo had done at the University of Pisa.

In 1718, Halley reported that three of the stars he was studying – Sirius, Arcturus and Aldebaran – had shifted over the centuries since Ptolemy. He strongly suspected that the discrepancies between the old charts and those of his own time were too large and too isolated to be attributed to errors in ancient measurement. Why should early astronomers have got every-thing else right but this? As a follow-up, Halley proceeded to measure the shift of Sirius during the 100 years since Tycho Brahe had observed it, a measurement that confirmed his suspicions. The change had been so gradual that it could only be noticed over a span of at least several human generations.

'Proper motion' is the name given to this movement of stars relative to one another over the centuries – an apparent movement across the sky when viewed from the Earth. Most stars are not moving only side-to-side, of course, as though the sky were a two-dimensional surface; they are likely at the same time to be getting closer to us or further away.

In 1720, at the age of 64, Halley succeeded Flamsteed as Astronomer Royal of England, which would not have pleased Flamsteed, for Halley had acquiesced in Newton's poor treatment of the old man. It probably would have pleased Flamsteed that his widow whisked all the instruments out of the Royal Observatory. They were legally hers because the financial arrangement at the Observatory was such that the Astronomer Royal purchased equipment out of his salary. Halley had to set to work acquiring new equipment.

Edmund Halley died in 1742 at the age of 85. One of his most significant achievements didn't come to fruition until nearly 20 years later.

When Halley had been on St Helena in his early twenties he had seen and timed a transit of Mercury across the Sun. He was the first man ever to observe both Mercury's first entry on to the Sun's disc and its final exit. Halley knew of a suggestion from James Gregory that a transit would provide an opportunity to use parallax in a new way to measure distances in the solar system. The passage of a planet across the Sun shows up as a tiny black dot passing across the Sun's face, and observers at different locations on the Earth's surface see the planet first touch the Sun's disc at different times.

Halley put little faith in earlier measurements of the Sun's and the planets' parallaxes and distances, including those of Cassini and Flamsteed, although his friend Newton came to agree with them. Newton also measured the orbits and the distances of the planets from the Sun, using not astronomical observations but the dynamics of the system as the basis for his calculation. He concluded that Cassini's and Flamsteed's results were better than his own. But as of 1700, the only real agreement among astronomers when it came to the Sun's distance was that it was at least 55 million miles away. Halley was convinced that the transit of Venus would be a chance to make much more definitive measurements.

A transit is a relatively rare event, but Halley knew there would be a transit of Venus across the Sun in 1761. He also knew he wouldn't be alive to witness it unless he lived to be 105. So he wrote and published detailed instructions on the best way to use observations of the transit from different parts of the world.

Sixteen years after Halley's death, with the return of the comet he'd seen in 1682, his name became a household word. In 1761 there was indeed considerable effort, much due to his prestige, to study the transit of Venus, and similar excitement about a second transit in 1769. Both times, the globe was studded with observing parties, who knew the opportunity wouldn't be repeated again until 1874. There are colourful stories connected with this venture which indicate that many astronomers at the time were less denizens of the ivory tower cum telescope than they were prototypes of Indiana Jones.

Frenchman Guillaume le Gentil planned to observe the 1761 transit from Pondicherry, near Madras in India. He arrived to find the town occupied by British forces. This was during the Seven Years War, England and France were enemies, and le Gentil was not welcome in Pondicherry. Rather than turn around and head for home, he settled nearby for eight years, supporting himself in part by trading while he waited for the next transit. By then the British had ceased to be an obstacle, but nature had no mercy on le Gentil. The Sun shone brightly before and after the transit. During the transit, alas, it was hidden by a cloud.

Jean d'Auteroche led another French observing team in Russia in 1761, and in what is now southern California in 1769. The party of four astronomers trekked overland across Mexico to reach their California observing location. D'Auteroche and two of the others died of disease shortly after their arrival. That left the fourth to undertake the treacherous return journey alone, but he brought back with him the dearly bought records of the observation.

The Reverend Nevil Maskelyne, sent by the Royal Society to St Helena to observe the 1761 transit, had a much better time of it. Maskelyne's expenditures were in the neighbourhood of £292, out of which £141 went on his personal liquor supply.

David Rittenhouse, in America, worked for months before the 1769 transit building a temporary log observatory at Norriton, near Philadelphia. He used a collection of instruments there that included telescopes from Europe and others he had built himself, and also his own eight-day clock that 'does not stop when wound up, beats dead seconds, and is kept in motion by a weight of five pounds'.

Charles Mason and Jeremiah Dixon, who would later establish the Mason-Dixon Line in North America, headed up another team sponsored by the Royal Society. That august body threatened them with disgrace and possible legal action if they failed to continue with their expedition to the Cape of Good Hope to observe the 1761 transit, after a French frigate attacked their ship in the English Channel and 11 crew members died. Evidently the captain of the French frigate had been unaware that in spite of the ongoing Seven Years War, Englishmen and Frenchmen were collaborating in this scientific endeavour.

Maximilian Hell, a Viennese Jesuit astronomer, observed the 1769 transit from Norway and suffered devastating damage to his reputation when Jerome Lalande insinuated that Hell had fiddled his observations to make them consistent with those reported by others. Karl von Littrow supported Lalande's allegation, claiming to have found proof in the form of different ink colours in Hell's report. Hell's good name wasn't restored until 1883, after there had been another transit. Among other things, it was discovered that von Littrow had been colour blind.

Sadly, the results of all this effort were less definitive than Halley and these astronomers had hoped. Precise determination of the instant the planet touched the Sun's disc was much more difficult than anticipated. Because of the Sun's corona and the atmosphere of Venus, Venus's image at the beginning and end of the transit was blurry. Calculations based on the results of these observations put the distance from the Earth to the Sun at about 95 million miles or 153 million kilometres, as compared with Cassini's measurement in 1672 of 87 million miles or 140 million kilometres, and our modern measurement of 93 million miles or 149.5 million kilometres.

Halley's discovery of proper motion was also destined to bear fruit far beyond his lifetime. It so happens that the three stars whose proper motion he first measured – Sirius, Arcturus and Aldebaran – are some of the brightest in the sky. Was this mere coincidence? A star might look brighter than others because it really is brighter, or it might look brighter because it is closer. Halley's discovery of proper motion gave astronomers a new clue.

Objects moving across our line of vision close to us appear to move more rapidly against the background than those further away. A child on a tricycle near us can easily outrace a car driving at a good clip off on the horizon. Logic tells us that the same will be true with stars that are moving across our line of vision. Unless stars are all the same distance, we ought to find the nearer stars appearing to move against a background of more distant stars. Most stars, in fact the vast majority of them, show no change of position since Ptolemy for a viewer on the Earth. Does that mean they are far more distant than those that have changed position?

Sixty-six years after Halley's discovery, astronomer William Herschel, the discoverer of the planet Uranus, studied the proper motions of a number of stars and the way those proper motions relate to one another, and from that he was able to plot the Sun's motion through our part of the Galaxy. Still later, the German astronomer Friedrich Wilhelm Bessel, suspecting that proper motion, rather than brightness, might be the most significant indicator of which stars are nearest us, used it as a basis for choosing which stars to try to measure with the parallax method.

Annual stellar parallax means the apparent shift in a star's position caused by our travelling from one extreme to the other of the Earth's orbit. As we return in our orbit to where we started observing, the star also returns to its original position.

Proper motion is the change in a star's position over a much longer period of time, due to the fact that all stars including the Sun move within the Galaxy. Proper motion was discovered more than a hundred years before anyone was able to detect annual stellar parallax.

The stellar parallax shift (see box above) that ancient Greek and Hellenistic astronomers could not find does exist, just as astronomers around 1700 were sure it must. Stars do have a parallax shift produced by the Earth's yearly journey around the Sun. But the shift is extremely tiny and difficult to detect. Certainly it isn't possible to see it with the naked eye, so the astronomers in antiquity can't be blamed for missing it. Telescopes of Galileo's and Cassini's time weren't refined enough to reveal it either.

One man who attempted to detect stellar parallax – making other important discoveries in the process – was James Bradley, born in England in 1693. The star Gamma Draconis passes almost directly overhead in London, and Bradley and his friend Samuel Molyneux, a wealthy amateur astronomer, decided to try to measure its parallax motion. They attached a 24-foot-long telescope to a stack of brick chimneys on the building where Molyneux was living. By using a screw, they could adjust the telescope to keep it tilted towards the star. The result was puzzling. Instead of having to adjust the tilt most in December and June, as they had expected, the adjustment was most extreme in March and September and was so large that, even if it had occurred at the right time of year, it was highly unlikely to be caused by parallax. Bradley took advantage of an exceptionally understanding aunt, who allowed him to cut holes in her roof and floors and install a larger and more sophisticated telescope. Observations with this instrument only repeated his and Molyneux's earlier baffling findings.

The explanation dawned on Bradley, so the story goes, while he was taking a cruise on the Thames. When the boat changed direction, a weathervane on the mast shifted. It wasn't the wind direction that had changed, however. It was the boat's direction in relation to the direction from which the wind was blowing. Bradley realized that the displacement of the stars he was studying was similarly caused by the changing motion of the Earth. Just as the wind direction seemed to shift according to the direction the boat was moving, so starlight seemed to shift according to the direction of the Earth's motion.

Bradley knew he had not found stellar parallax. What his observations demonstrated was that the Earth orbits the Sun and the speed of light is not infinite, both of which were already well accepted. Bradley named the effect he had found 'aberra-

1. A 15th century woodcarving of Ptolemy (second century AD), in the Ulm Cathedral: 'When I trace at my pleasure the windings to and fro of the heavenly bodies, I take my fill of ambrosia, food of the gods.'

2. Drawing from Nicolaus Copernicus's book *De revolutionibus* (1543), placing the Sun at the centre with the planets orbiting it. Beyond them is the 'immobile sphere of the fixed stars'. The Moon orbits the Earth.

3. Johannes Kepler (1571–1630), discoverer of eliptical orbits: 'Don't sentence me completely to the treadmill of mathematical calculations - leave me time for philosophical speculations, my sole delight.'

4. Aristotle, Ptolemy and Copernicus on the Frontispiece of Galileo's *Dialogo* (1632). The title page assures that this 'discourse concerning the two chief systems of the world, Ptolemaic and Copernican,' discusses 'without prejudice one view, then the other, on the basis of philosophy and natural law'.

DIALOGO

D I

GALILEO GALILEI LINCEO
MATEMATICO SOPRAORDINARIO
DELLO STVDIO DI PISA.
E Filosofo, è Matematico primario del
SERENISSIMO

GR.DVCA DI TOSCANA.

Doue ne i congressi di quattro giornate si discorre
sopra i due

MASSIMI SISTEMI DEL MONDO
TOLEMAICO, E COPERNICANO;

*Proponendo indeterminatamente le ragioni Filosofiche, e Naturali
tanto per l'vna, quanto per l'altra parte.*

CON PRI VILEGI.

IN FIORENZA, Per Gio:Batista Landini MDCXXXII.
CON LICENZA DE' SVPERIORI.

5. Woodcut showing the Octagon Room at the Royal Observatory in Greenwich as it looked in Flamsteed's time, in the 17th and early 18th century.

6. One of the earliest photographs ever taken: Sir John Herschel's 1839 photo of his father's great 40-foot telescope.

7. Edmond Halley (1656–1742): 'There remains but one observation by which one can resolve the problem of the Sun's distance, and that advantage is reserved for astronomers of the following century.' He left instructions for studying the transit of Venus across the Sun in 1761.

8. Sir William Herschel, astronomer and composer (1738–1822): 'I used frequently to run from the harpsichord at the theatre to look at the stars during the time of an Act and return to the next Music.'

9. Caroline Lucretia Herschel, in a drawing made when she was 97, in 1847. A friend wrote that it does not 'do justice to her intelligent countenance'.

10. The 'Leviathan of Parsonstown', built by Lord Rosse in the 1840s and through which he observed the spiral nebulae. It remained for more than half a century the largest telescope in the world.

11. Henrietta Swan Leavitt (1868–1921). Her discovery of Cepheid variable stars and how to use them as stellar yardsticks led to the first measurements of distances outside the Milky Way Galaxy.

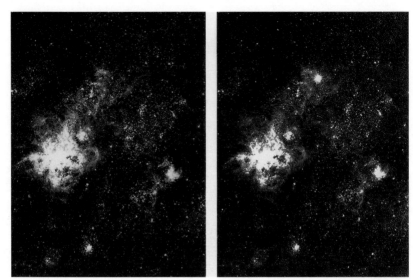

12. Before and after photos of Supernova 1987A in the Large Magellanic Cloud. The image on the left was taken in 1969; that on the right in February 1987, about a week after the supernova appeared.

13. The Andromeda galaxy, dominant spiral galaxy of the Local Group, with two companion galaxies – M32 (the bright dot close to Andromeda on the left) and NGC 205 (the bright dot on the right).

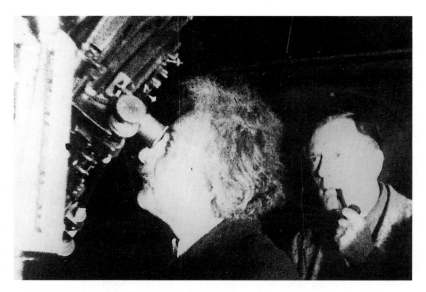

14. Albert Einstein visits Edwin Hubble at Mt. Wilson in 1930. Hubble's astronomical observations and Einstein's theories both indicated that the universe must be expanding.

15. Grote Reber's radio telescope in Wheaton, Illinois (1937). Reber spent $4,000 of his own savings on designing and building this device in his mother's back yard.

16. This image of the Milky Way Galaxy in 'near infrared' wavebands, from observations of the Cosmic Background Explorer (COBE), clearly shows the thin, flattened disk and central bulge.

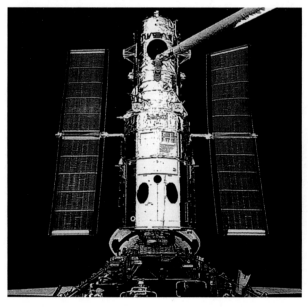

17. The Hubble Space Telescope, NASA's orbiting observatory, linked to the Space Shuttle Endeavour during the December 1993 repair mission.

tion' and announced the discovery to the Royal Society in 1729. Aberration produces a 20½ arcsecond shift (see Figure 4.4) in the apparent positions of stars over a year. Bradley also found that the Earth wobbles due to the fact that its shape is not perfectly spherical, and he gave the wobble the name 'nutation'. Aberration and nutation were not what Bradley had set out to find; but these discoveries were actually helpful steps on the road to discovering the tiny displacement of true annual stellar

Figure 4.4

The arcsecond originated as an ancient Mesopotamian measurement. A circle has 360 degrees. Each degree can be divided into 60 minutes of arc or 'arcminutes'. Each minute of arc can be divided into 60 seconds of arc or 'arcseconds'.

The measurement of an arcsecond isn't a measurement of true size. If you hold your finger up at arm's length against the sky, its width covers about two degrees of arc. But this finger held at arm's length would cover a branch on a nearby tree, the Concorde flying overhead, the entire Moon, or (out of sight with the naked eye) an enormous number of galaxies. Clearly not all objects in the sky that have the 'angular size' of two degrees of arc are the same true size. How 'large' two degrees of arc are depends upon how deep into space you're looking. The Moon's angular size is about half a degree of arc. The Sun's angular size is approximately the same as the Moon. Yet these two bodies are definitely not the same true size.

In the picture below, each of the circles has the same angular size viewed from Earth, covering the same number of degrees of arc, yet they are not the same true size. (Recall Aristarchus's study of the Sun and Moon, shown in Figure 1.4.)

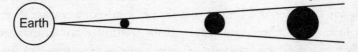

One arcsecond is the angular size the width of your finger would have if you were able to hold your finger up about 5,000 feet (1,500 metres) above you.

parallax. In any search for annual stellar parallax, one needed to take into account these other reasons why the positions of stars change with the seasons. The negative side of Bradley's discoveries was even more significant. He had shown that the parallaxes of stars could not be more than one second of arc. Had they been as large as one arcsecond, he knew he would have been able to detect them. This meant stars were much further away than was generally supposed.

In 1742, Bradley succeeded Halley as Astronomer Royal.

The 18th century saw a rapid increase in the number of observatories in Europe. Among the sciences, only in medicine were there more people professionally involved in research. Not all of them were peering through telescopes, for the meticulous cataloguing of what others found required many clerical workers. Funding came from governments, universities, scientific bodies, even religious societies – the expense often justified in terms of the practical spin-off for navigation, mapping and surveying, and the prestige such institutions gave to their sponsors.

The educated public loved astronomy and followed it avidly. Travelling lecturers drew large audiences, less technical books were best-sellers, telescopes for amateur use sold well, as did globes. Natural theology – the argument that nature and the harmony of the universe are eloquent proof of the existence of God and tell us what God is like – was preached from many pulpits, given voice in hymns and secular poetry, and favourably discussed in academic circles. Though astronomy had ceased to be a required part of the curriculum in most universities, it was widely considered essential to a proper gentlemanly education.

In England, the Royal Observatory continued to be financed from government coffers, a considerable investment. The Observatory's demand for equipment as well as the Royal Society's great interest in the improvement of all scientific instruments, including telescopes, helped support and encourage a healthy local optical industry. The Industrial Revolution, which started in England in the 18th century, brought advances in the design and construction of machines and in precision engineering in general. Astronomy reaped enormous benefits from this progress and also, with the increasing expertise of its instrument-makers and demand for their products, contributed substantially to it. The finest telescopic instruments came from England, and English manufacturers were suppliers to all Europe, setting the standard and style of the profession. Earlier astronomers had often designed and built their own instruments, and some of the more notable among them would continue to do so, but increasingly there began to be a division between those who manufactured telescopes and those who used them. Telescope builders were not necessarily considered a lower breed than practising astronomers. Some were elected to the Royal Society.

In the early 19th century, the Industrial Revolution spread to Europe and the United States. London opticians were still producing reflecting telescopes, the type pioneered by Newton and others in the late 17th century (see Figure 4.5), at affordable prices. English amateurs were making some excellent instruments themselves. However, Britain's Prime Minister William Pitt and his government dealt London's optical industry an almost mortal blow by imposing a punitive tax first on windows and a little later on all glass. After Swiss glass-maker Pierre Guinand moved from Switzerland to Munich in 1804, bringing with him a new method of mixing molten glass, Germany took the lead in refining the art of telescope design and manufacture. Bavarian-born Josef von Fraunhofer, who had worked for a time as Guinand's assistant, greatly improved the refracting telescope (again, see Figure 4.5). Friedrich Wilhelm Bessel pioneered the use of the meridian circle, which made it possible to measure two coordinates of a star at the same time, improving greatly the accuracy of observations, as did advances in the design of astronomical clocks. Mathematician Karl

Figure 4.5

This sketch shows the basic difference between a refracting and a reflecting telescope. There are many varieties of each.

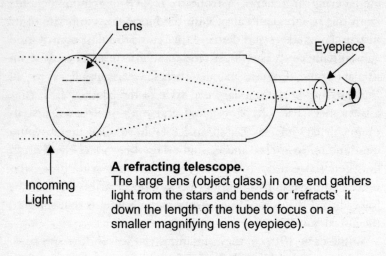

Lens

Eyepiece

Incoming Light

A refracting telescope.
The large lens (object glass) in one end gathers light from the stars and bends or 'refracts' it down the length of the tube to focus on a smaller magnifying lens (eyepiece).

A reflecting telescope.
A curved mirror at the bottom end of an empty tube gathers light from the stars and reflects it to focus on a second mirror suspended at the first mirror's focal point. The second mirror reflects the image to the eyepiece.

Eyepiece

Incoming Light

Mirror

Curved Mirror

It's possible to make a reflecting telescope much larger than a refracting telescope, because the large mirror can be supported from the back as a lens cannot.

Friedrich Gauss, then only 18 years old, at Göttingen in 1804, invented the method of 'least squares' which enabled an astronomer to choose the best observations in a less arbitrary manner.

Finally, by the late 1830s, improved technology and theoretical understanding had converged to the extent that it was possible for three astronomers to detect annual stellar parallax. Clearly it was a discovery whose time had come. The three measurements occurred independently but almost simultaneously.

Friedrich Wilhelm Bessel, in Königsburg, Germany, was the first to announce his findings, in 1838. Reasoning that proper motion, rather than brightness, might be the most significant indicator of which stars are nearest, he chose 61 Cygni, a dim star with a large proper motion (5.2 arcseconds a year). (See Figure 4.4.) The figure at which he arrived for its annual parallax was 0.3136 arcseconds. Knowing the distance the Earth had travelled in its orbit in order to produce this displacement allowed him to calculate the distance from the Earth to the star – 3.4 parsecs (11.2 light years) – which is, for comparison, 600,000 times greater than the distance from the Earth to the Sun. Bessel's measurement was a considerable step up on the cosmic distance ladder.

Meanwhile Thomas Henderson from Scotland, observing from South Africa, chose to study Alpha Centauri. He made his choice on the basis of brightness rather than proper motion. Alpha Centauri is the third brightest star in the night sky. Though Henderson actually measured Alpha Centauri's parallax before Bessel measured 61 Cygni's, Henderson didn't announce his results until he got back to Britain early in 1839 – thus losing out to Bessel in the pages of history. A year later, Friedrich von Struve, born in Germany but working in Tallin (now in Estonia), announced that he had measured the parallax for Alpha Lyrae (aka Vega), the fifth brightest star in the night sky.

The parallaxes measured for all three of these stars were small:

For 61 Cygni, a parallax of 0.3136 arcseconds (see Figure 4.4), a distance from us of 3.4 parsecs (11.2 light years). We now know that 61 Cygni is a double star.

For Alpha Centauri, about 1 arcsecond parallax – a figure later refined to 0.76. The Alpha Centauri *star system* (for we now know it consists of three stars) is 1.3 parsecs away (4.3 light years). The star that we call Proxima Centauri is one of that trio and at the moment is the solar system's closest neighbour. It revolves around its mates Centauri A and B every 500,000 years.

For Alpha Lyrae (Vega), a parallax of 0.2613 arcseconds, a distance of 8.3 parsecs (26 light years).

How far away they are! And yet these are some of the nearest stars. With their measurement, it began to sink home how profoundly alone our little solar system community is. After Cassini and Flamsteed it had seemed so huge. Now it became tiny compared to the enormous emptiness we would have to cross to reach anything else beyond. You have to multiply the distance from the Sun to Pluto (the planet furthest from the Sun) by nine thousand to reach Proxima Centauri – the next break in the darkness.

The successful measurement of the distance to the nearest stars strengthened an impression that had existed since the 18th century that 'celestial mechanics', the marriage of mathematics and astronomy, was the highest of all the sciences and the most valuable for deepening human understanding of the laws of nature. Another achievement crowned this reputation even more spectacularly. Soon after Bessel, Henderson and von Struve's measurements, Urbain Jean Joseph Leverrier, who was the virtual dictator of the Royal Observatory in Paris and scorned and discouraged any intellectual pursuit that *wasn't* celestial mechanics, studied the orbit of the planet Uranus. He came to the conclusion that certain mysterious irregularities in the orbit that do not accord with Newton's laws must be caused by the gravitational pull of another undiscovered planet. On 23 September 1846, Johann Gottfried Galle at the Berlin Observatory, looking for that unknown planet where Leverrier had predicted it should be, and working with records kept by British astronomer John Couch Adams, discovered Neptune. Leverrier's prediction had been astoundingly close to right – quite by coincidence, actually, for he had not chosen correctly among several possible solutions for the orbit. The discovery was a public sensation. A miracle! Indeed, more of a miracle than the public knew, given Leverrier's wrong choice.

At mid-century, astronomy and celestial mechanics did indeed seem to be moving from triumph to triumph. They had also suffered one setback, for knowing the distances to a few stars gave astronomers a way to calculate their absolute magnitudes and erased forever any hope that the absolute magnitude of all stars is the same. The elephants on the plain came in a variety of sizes, and we could measure directly the distance to only a few of them. How could we possibly judge the size or distance of the others?

One way to approach the problem would be to take another look at this category we've been calling 'elephant' and see whether we can break it down. Maybe there are Indian elephants and African elephants, and some way to tell them apart. If all elephants aren't the same size, perhaps all Indian elephants *are*.

The trick would be to find characteristics that (unlike apparent size) won't change with distance, such as distinctive ears. We could call these strange-eared animals Group A elephants. Suppose we do have a way of measuring the actual distance to a few of the Group A elephants and having done so discover that their size doesn't vary greatly. It seems fairly safe to assume that those Group A elephants too far away for direct measurement are also that size. With that assumption, we can take the exercise further. If another animal is standing near a

Group A elephant in the distance, drinking from the same waterhole, we can judge the size of that second animal by comparing it to the elephant. Suppose the second animal is spotted and has an extraordinarily long neck. We name it a giraffe. Now if we go on watching elephants and giraffes out there and conclude that all giraffes are about the same size, we have a potential way of calculating the distance to any other animal that shares giraffe characteristics. One measurement builds on another. Of course if there is a mistake somewhere along the line – maybe Group A elephants come in two sizes, or maybe the animal we think is near the elephant is actually fifty feet beyond it – then the whole measurement structure begins to collapse and has to be recalibrated.

Even before the first stellar parallax measurements, astronomers had begun to hope that, upon closer scrutiny, stars would turn out to have different characteristics that would allow them to be grouped into categories or 'families'. If stars don't all share the same absolute magnitude, perhaps those within certain recognizable 'families' do.

There had been some developments that would lead to a better understanding of stars. At the beginning of the 19th century, it had been generally assumed that it would never be possible to discover the chemical composition of stars or their physical make-up, because researchers couldn't get near enough to examine them. The French philosopher Auguste Comte pointed to the chemical composition of stars as an example of 'unobtainable knowledge'. Not everyone shared this pessimism. Researchers were soon to find that starlight carries with it an enormous amount of information about its source, if you can crack the code.

Since Isaac Newton's study of optics, scientists and the general public had known how to use a glass prism to break a ray of light into its component parts. When white light passes through the prism, the colours of which the light is composed spread out in an ordered sequence – the spectrum – the familiar rainbow. The order is always the same: red, orange, yellow, green, blue, indigo and violet. The acronym for that is 'Roy G. Biv'.

We refer to position in the spectrum by colours ('the red end of the spectrum' or 'the violet end of the spectrum'), or more precisely by wavelengths, because each colour is

Figure 4.6   The Electromagnetic Spectrum

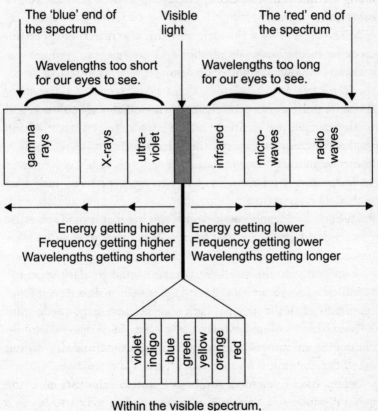

The 'blue' end of the spectrum

Visible light

The 'red' end of the spectrum

Wavelengths too short for our eyes to see.

Wavelengths too long for our eyes to see.

gamma rays

X-rays

ultra-violet

infrared

micro-waves

radio waves

Energy getting higher
Frequency getting higher
Wavelengths getting shorter

Energy getting lower
Frequency getting lower
Wavelengths getting longer

violet
indigo
blue
green
yellow
orange
red

Within the visible spectrum, our eyes sense different wavelengths as different colours.

produced by a different wavelength of light. The longer waves are the red. The waves grow shorter as we move across the spectrum to violet. See Figure 4.6.

Light that human eyes can see – the visible spectrum – is only a small part of the much larger 'electromagnetic spectrum'. What is out beyond red on the one hand and violet on the other is invisible to us, but there is a great deal out there – infrared light, ultraviolet light, gamma rays, X-rays and radio waves, all of them forms of electromagnetic radiation, with wavelengths either too short or too long to be within the visible spectrum.

When light passes through a prism, the resulting spectrum gives us information about the light source, even when that source is billions of light years away.

- An incandescent, solid light-source radiates all colours, and the spectrum is continuous from violet to red (and beyond the visible spectrum in either direction).
- An incandescent gas radiates only a few isolated colours and each different kind of gas has its own pattern, called an emission spectrum.
- When an incandescent solid (or its equivalent) is surrounded by a cooler gas, the result is a spectrum in which a continuous background (such as the spectrum an incandescent solid would produce) is interrupted by dark spaces – called 'absorption lines'. See Figure 4.7. In this case the gas surrounding the original light source has absorbed from that light those colours which the gas would radiate itself. By looking at the pattern of the absorption lines and noting where they fall within the spectrum, it's possible to discern which gas or gases are responsible for the absorption.

Much of our understanding of light and spectra stems from the pioneering work of Josef von Fraunhofer, born in Staubing, Bavaria, in 1787. Von Fraunhofer was the eleventh and youngest

Figure 4.7 Absorption Spectra

When an incandescent solid is surrounded by a cooler gas, the result is a spectrum in which a continuous background is interrupted by dark spaces – called 'absorption lines' – that occur because the gas has absorbed from the light those colours which the gas would radiate itself.

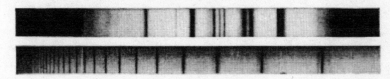

child of a master glazier and worker in decorative glass. Orphaned at 12, he became the apprentice of a mirror-maker and glass-cutter in Munich who paid him nothing, offered minimal instruction, and made it impossible for him to attend the Sunday Holiday School which offered apprentices a little schooling outside their trade. Fraunhofer's luck turned for the better when he was 14 and his master's house collapsed, burying the boy in the ruins. His escape – he was injured but protected from death by a crossbeam – became a news item in Munich and reached the ears of the Elector Maximilian. Maximilian, touched, gave young Fraunhofer some money which he used wisely, purchasing a little equipment for himself and buying out of his apprenticeship. He had to return only temporarily when his own business (engraving visiting cards) failed to support him. Fraunhofer's miraculous survival in the collapsing house also drew the attention of a wealthy Munich lawyer and financier named Utzschneider, who soon hired Fraunhofer to work at his glass-making establishment. Such was Fraunhofer's innate ability and zeal for his craft that when he was in his early twenties Utzschneider had already put him in sole charge of the glassworks.

Fraunhofer was one of a handful of men in the early 19th century who rose from working-class backgrounds to become leaders in astronomy. In a short lifetime, he designed and built increasingly fine telescopes, among the best in the world at that time, and he was responsible for a number of inventions that made their use more effective. Bessel and von Struve were using Fraunhofer telescopes when they first measured stellar parallax.

Fraunhofer's discoveries about light led to some of the most significant developments of the 19th and 20th centuries, making him one of the most important figures in the history of optics. He was the first to study and map the absorption lines of the Sun's spectrum.

In 1814, Fraunhofer was trying to find ways to make more accurate measurements of the way a piece of glass refracts the light. When Isaac Newton had studied the spectrum of light, he'd done so by allowing sunlight entering a round hole in a shutter to pass through a glass prism and fall on a screen. Fraunhofer used a modification of Newton's experiment. Basically, what he did was to substitute a narrow slit for the round hole and a telescope for the screen.

Fraunhofer found that the continuous spectrum of the Sun is interrupted by many dark lines, and he found the lines present in *all* sunlight, whether direct or reflected from other objects on Earth or from the Moon and planets. He labelled the 10 strongest lines in the solar spectrum and recorded 574 fainter lines. See Figure 4.8.

Continuing to investigate, Fraunhofer found these lines also appearing in the spectra of stars, but in different arrangements. He concluded that the lines must originate in the very nature of the Sun and the stars. What he had actually discovered was the signature of the different chemical elements present in the Sun's and stars' atmospheres, the third type of spectrum we described above, an 'absorption spectrum', the type in which the cooler gas surrounding the original light source has absorbed from that light those colours which the gas would radiate itself. Fraunhofer came closest to realizing these implications when he noticed that

Figure 4.8

Fraunhofer's map of the solar spectrum. He used the letters to identify the most prominent absorption lines.

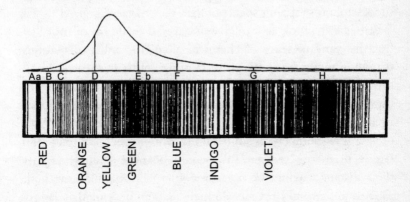

two dark lines in the Sun's spectrum coincided with two bright lines in the spectrum of his sodium lamp.

Fraunhofer died of tuberculosis at the age of 39. Not long before that he was knighted and relieved of his taxes as a citizen of Munich.

In 1849, W. A. Miller at King's College, London, and Léon Foucault in Paris independently recognized that the pair of lines Fraunhofer had found in the spectrum of the Sun – and that had matched the two lines in the spectrum of his sodium lamp – were present in the spectrum of sodium in the laboratory. Ten years later, Gustav Kirchhoff and Robert Bunsen were able to explain the full significance of that discovery: the Sun is an incandescent body surrounded by a gaseous atmosphere with a lower temperature. The lines mean that sodium is present in the atmosphere of the Sun. In the early 1860s, William Huggins, a British astronomer whose wealth allowed him to construct a private observatory, and Miller found that light coming from other stars had the same spectral lines as light from the Sun. A new era of astronomy had begun. From 1863 to 1868, Father Secchi at the Vatican provided the foundations of the classification of stars by the patterns of their spectra. Also in the 1860s Huggins in England and Draper in America began to have success photographing spectra.

It began to look as though stars are all made up of more or less the same mixture of chemical elements, and expectations that they might be sorted into families by their spectra dwindled. But further investigation showed that certain patterns of lines in the spectra are not alike for all stars after all. Hopes rising again, astronomers began to try to categorize stars. With luck, a few members of some family whose spectrum made the family members recognizable would be near enough so that their distances could be measured using the parallax method. This sample would reveal whether all members of that family shared the same absolute magnitude and whether they could be used as distance calibrators.

In the 1840s there was another development that would have dramatic long-range benefits for astronomy and astrophysics. Austrian physicist Christian Doppler discovered what we now call the 'Doppler effect'. In everyday life we experience it with sound rather than light, as the drop in the pitch of a fire engine siren when the vehicle approaches us, passes, and moves away. As it approaches, sound waves coming to us from it are bunched up, shortened. As it moves away from us, sound waves from it are stretched out. In either case, the sound waves are 'shifted' from the length they would be if the source were standing still. Our ears interpret the lengths of sound waves as different pitches, the longer the wave the lower the pitch.

Doppler thought the same effect would occur with light, which likewise can be thought of in terms of waves and wavelengths, and in 1848 French physicist Armand Fizeau was the first successfully to describe 'red shift' and 'blue shift' in light. Today we use the term 'Doppler shift' not only for sound, but also for light and all other radiation in the electromagnetic spectrum.

Just as our ears interpret sound waves of different lengths as different pitches, our eyes interpret light waves of different lengths as different colours, the longer the light waves the nearer the 'red' end of the spectrum. (Review Figure 4.6 above.) Because of the Doppler effect, an observer finds the light from a star that's receding shifted towards the red end of the spectrum, a red shift; and the light from a star that's approaching shifted towards the blue end of the spectrum, a blue shift. For some reason, in scientific parlance, indigo and violet are ignored.

Doppler at first believed that red and blue shifts were causing the differences of colour observed in some binary stars, but other researchers, including Fizeau, soon pointed out that the shift can't be seen at all as a visible difference in colour. Instead, the shift shows up as a slight yet measurable shift of the spectral lines in the light coming from a star.

The discovery that the pattern of lines in the spectrum of sodium in the lab is also present in the spectrum of light from the Sun had taught researchers that the Sun contains sodium. Measuring the positions at which such familiar lines appear in the spectrum of a star, and comparing these with the wavelengths at which the same lines appear in light in a laboratory, reveals whether the pattern has shifted. If the pattern has shifted towards the red end of the spectrum, the star is moving away. If it has shifted towards the blue end, the star is moving towards us. The amount of shift can be directly related to the speed at which the star is approaching or receding. Huggins was the first to determine the velocity of recession of a star.

Though the Doppler shift was obviously a remarkable help in plotting the movement of stars, astronomers also realized that a star's motion is more complicated than the simple velocity of recession the shift reveals. The Doppler shift by itself is only the measure of how rapidly a star is increasing or decreasing its distance from us, not what direction it's taking or even how fast it's actually moving. Only a few stars move directly *along* our line of sight (directly away from or towards us). Most are veering off at an angle, and that makes it much more difficult to determine a star's speed and overall motion. For example, a star moving very rapidly *across* our line of sight isn't increasing or decreasing its distance and would show no shift at all.

Related to this problem, there is a method that proved so effective for not-too-remote groups of stars that some of the measurements it provided weren't bettered until well after the advent of space-based telescopes. The technique is in fact still used for more distant groups. Here was the problem to be solved: For a single star, it was at best possible to determine: (1) its apparent motion across our line of sight, measured in arcseconds; and (2) its recession/approach rate, calculated from its red shift. This rate is an actual speed in miles or kilometres per second, while the measurement in arcseconds of the star's motion across our line of sight is not. (Review Figure 4.4 on p. 131.) An arcsecond, unlike a mile or a kilometre, isn't an absolute distance and can't be transformed into one unless we know how far away the star is.

The challenge then is to calculate in what manner the two measurements (recession/approach rate and motion across our line of sight) combine to give the actual 'speedometer' speed in miles or kilometres, and this is where the advantage of having a group of stars, rather than a single star, comes in. If a cluster of stars is moving through space, it's reasonable to assume that the stars are all travelling in nearly parallel lines and that the effect of perspective will cause these paths to seem to draw together towards a single point in the sky. We see a similar effect if we watch the two parallel rails of a railway track from the last carriage in the train. The rails seem to meet in the distance. For a cluster moving away from or toward us, the effect shows up if we watch over a period of many years. The stars look as though they're getting closer together or further apart. That is, their paths appear to converge or diverge.

Studying this pattern – the way a star cluster seems to shrink or swell, and the location of the point where the stars' paths appear to meet – astronomers can determine whether a cluster is moving directly along our line of sight or, if not, at what angle to our line of sight it is travelling. That angle tells them what proportion of the stars' true (speedometer) movement must be side-to-side and what line-of-sight. Knowing that proportion and the recession/approach rate, it's possible to calculate a number, stated in miles or kilometres per hour, for the stars' motion across our line of sight. It then becomes an answerable question of mathematics: How far away does a group of stars have to be for *this* velocity to produce *this* many arcseconds of shift across the sky in one year?

This technique is called the 'moving cluster method'. It provided the distance to the nearest star cluster, the Hyades, which forms the head of Taurus, the bull. The Hyades lie some 40 parsecs away (approximately 140 light years), about 10 parsecs further out than the parallax method alone could measure in the late 19th century. Fortunately, the Hyades include many different kinds of stars. Knowledge of their distance allowed experts to calculate the absolute magnitude of stars in the cluster whose spectral lines identified them with particular families of stars. From there, the technique of comparing the apparent magnitude of stars belonging to the same family sufficed to reveal approximate distances to stars and clusters much further away.

Another technique that became possible with the discovery of how to measure the red shift of a star is called 'statistical parallax'. It contributed directly to one of the first major advances in measurement that took place after the turn of the century. Astronomers choose a large group of stars, basing their choice on some common characteristic such as colour or spectrum. They measure their red or blue shifts to learn their velocities along our line of sight and then find the average of these velocities for all the stars in the group. The assumption, a reasonable one, is that stars are moving every which way in the sky at many speeds and in many directions and that in a large enough group of stars all this motion averages out. So it also seems safe to assume that the average velocity along our line of sight is also the average velocity *across* our line of sight. Again, we are prepared to ask the question: How far away do stars have to be for *this* velocity to produce *this* many arcseconds of shift across the sky in one year? The result now is the *average* distance for the stars in the chosen group.

That might seem to tell us very little. However, with some refinements, this was the method that would later provide distances to some Cepheid variable stars, and these would provide a bridge to objects many, many light years further away.

Along with increased ability to measure the distances to stars and groups of stars went a deepening curiosity about the larger picture – how everything fits together. What had all this burgeoning knowledge been leading men and women to believe about the overall shape, structure and size of the universe? To answer that, we must return to the 18th century.

CHAPTER 5

Upscale Architecture
1750–1958

The subject of the Construction of the Heavens, on which I have so lately ventured to deliver my thoughts to this Society, is of so extensive and important a nature, that we cannot exert too much attention in our endeavours to throw all possible light upon it.

William Herschel, 1785

In the decades following Galileo's observation that the Milky Way is made up of myriads of stars, few astronomers followed up on his discovery. There was plenty to keep them occupied within the solar system. Not until the middle of the 18th century is there a record of anyone suggesting that the stars also might be arranged in some sort of 'system' – or system*s*. Even then it was not astronomers who introduced the idea. There was no observational evidence to support it. We are told that the French court thought about it and found it an interesting topic of conversation. No one otherwise recorded their conclusions.

Thomas Wright of Durham, England, who called himself a philosopher, not an astronomer, was one of the first to propose a larger arrangement. Wright spent most of his teenage years, while apprenticed to a clock-maker, immersed in astronomy. His father found that so distasteful that he burned young Thomas's astronomy books. Later, Wright became a sailor (giving that up after a storm on his first voyage), a mathematics tutor (until he became involved in a scandal with a clergyman's daughter), a navigation instructor for sailors, a surveyor, a successful teacher and consultant among the aristocracy on the subjects of philosophy and mathematics, and finally an author. All this he managed to do in spite of having a speech impediment.

In a book entitled *An Original Theory or New Hypothesis of the Universe*, published in 1750, Wright suggested that the Milky Way, which he called the 'universe' or the 'creation', was a slab of stars whose centre was a supernatural source of energy, goodness, morality and wisdom. He thought that this slab might be one among many 'creations' of its kind and that faint clouds of light, known as the nebulae, might be the other 'creations'.

The great East Prussian philosopher Immanuel Kant, who was a trained mathematician, read an account of Wright's ideas in a newspaper and was sufficiently taken with them to try to give them a more mathematical and scientific footing. Kant was not an experimental scientist, nor did he observe the heavens through telescopes. Instead he ruminated about the meaning of the observations and discoveries others were making. He agreed with Wright that if the Milky Way is a flat slab of stars, then other fuzzy patches in the sky must be similar flat slabs. Our slab must be one of many in an enormous universe.

One man who owned a copy of Thomas Wright's book was William Herschel, a prominent professional musician and composer who'd become enamoured of astronomy as a child. Born in Hanover, Germany, in 1738, Herschel first visited England when he was 19 as a member of the band of the Hanoverian Guard. Within the year he had resigned his commission and moved to England to pursue a musical career there, where he became Director of Music for the City of Bath and organist at the Octagon Chapel. Herschel spent most of his time teaching music students, engaging artists for concerts, and composing. His works include 24 symphonies, seven violin concertos and two organ concertos, and he also wrote 'glees and catches' and anthems for his choirs. Yet history remembers him as an astronomer, for at the age of 35, during hours when others were sleeping and whenever his pupils went away on holiday, he turned again seriously to his boyhood hobby.

In 1781, scanning the heavens with a telescope along with his younger sister Caroline, looking for faint, distant objects, Herschel discovered the planet Uranus. As a result of that, George III appointed him his personal astronomer, with few duties other than to keep the royal family advised on the subject of astronomy and any new wonders in the skies, and occasionally to impress visiting dignitaries. In 1787, the king also granted Caroline a salary.

The Herschels moved to Datchet, near Windsor, so that William could perform his new duties more conveniently. To supplement what was not a large stipend (less than he earned as an organist) he began to make and sell telescopes, and he also set about building bigger and better instruments for his own use. Herschel's telescopes were reflecting telescopes, a type that had fallen somewhat out of favour. Herschel's success gave them a new popularity. (Review Figure 4.5.) His largest was a 40-foot instrument with a 48-inch mirror. Its tube was five feet in diameter. Herschel, still the musician, held a small concert in the tube to celebrate its dedication.

Although the size of the 40-foot telescope made it a curiosity and a source of pride for Herschel and for the king, it really was something of a liability. So ungainly and unsafe was this behemoth that King George's workmen were reluctant to help manoeuvre it. What was worse, Herschel found himself wasting precious time and good viewing nights (rare in England) explaining and demonstrating this wonder to sightseers, who included the king and his family and royalty and astronomers from all over the world. 'Come, let me show you the way to heaven,' the king was overheard murmuring to the Archbishop of Canterbury, as he ushered him towards the telescope.

Herschel ended up preferring his more manageable 20-foot reflector, though the use of that also involved physical risks. His sister Caroline wrote:

> My brother began his series of sweeps when the instrument was yet in a very unfinished state, and my feelings were not very comfortable when every moment I was alarmed by a crack or fall, knowing him to be elevated fifteen feet or more on a temporary cross-beam instead of a safe gallery. The ladders had not even their braces at the bottom; and one night, in a very high wind, he had hardly touched the ground before the whole apparatus came down.

Just before Herschel's 40-foot telescope was dismantled, 17 years after his death, his son Sir John Herschel composed a ballad, and he and his family went into the tube on New Year's Eve to sing it. The great telescope's life ended as it had begun, with a concert rather than an astronomical observation.

For William Herschel, building bigger telescopes was not primarily a matter of hubris or impressing the king. He wanted to learn how the universe is constructed on the largest scale. Rather than tackle the whole at once, Herschel chose 700 different regions of the sky on which to concentrate his attention. In each of these regions he proceeded meticulously to count the stars of different brightnesses and to catalogue all the double stars he could find, for he hoped to measure their annual parallaxes. Thus began the first project to map the universe in three dimensions, something astronomers are still doing 300 years later.

Herschel's map turned out to resemble the late 20th-century model of the Galaxy, which is remarkable because he, as

Newton had done when he measured his star distances, worked from unsound assumptions. Herschel allowed himself to assume that all stars have essentially the same absolute magnitude. He chose the star Sirius, the brightest star in the night sky, as his standard. That is, he proceeded on the assumption that if all stars were the same distance as Sirius, they would all look just as bright, so the extent to which they appear to differ in brightness from Sirius can be used as a gauge of their distances. Herschel can't be as easily excused as Newton for this error, because he must have been aware of a strong argument coming from his contemporary John Michell. Michell, a natural philosopher best remembered as the first to suggest the existence of 'dark stars' – what we now call black holes – pointed out that the stars of the Pleiades definitely do not all appear equally bright, and yet, grouped as they are, they must surely be about equal in distance from us.

Herschel also decided to assume that all stars are evenly distributed and that through his telescope he was seeing all the way to the outermost regions of the star-filled universe, so that his star counts really were accurate indications of the total number of stars.

Herschel's map had the arrangement of stars as a spiky, flat, elongated blob, a model that got the nickname of the 'grindstone', though it is difficult to imagine that any grindstone could be so ill-designed and jagged. (See Figure 5.1.) The spiky edges were the result of dark rifts that Herschel observed in the Milky Way. He thought these were probably holes in space and that through them he was seeing emptiness beyond.

Herschel was so bold as to estimate the size of the grindstone. He had used the star Sirius as his standard, so he decided to call the distance to Sirius, whatever it might turn out to be in miles or kilometres, one 'siriometer'. Herschel calculated that the grindstone measured 1,000 siriometers across and was 100 siriometers thick. When he made his estimate, the first detection of stellar parallax was still 50 years in the future, but it's

Figure 5.1

William Herschel's map – the 'grindstone' – was surprisingly like the modern picture of the Milky Way Galaxy.

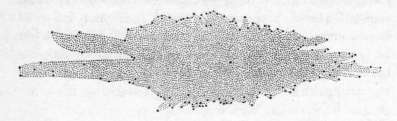

now possible to attach definite numbers to his scheme, because Sirius turns out to be not quite nine light years away. That would make Herschel's grindstone about 9,000 light years from end to end and 900 light years thick. Modern estimates of the dimensions of the Galaxy have it more than 10 times that large.

William and Caroline Herschel also scrutinized the nebulae. The big question in the 1780s was, are they clusters of many stars? The Herschels had the best equipment in the world for finding out. William soon reported with delight that many nebulae *were* resolvable into stars. He even thought that he could *almost* make out individual stars in the Andromeda nebula, though astronomers now are sure that would have been impossible with his telescopes. In 1790 the Herschels confirmed the existence of another kind of nebula in which a cloud of luminous gas surrounded a single central star. Herschel thought this might be a planetary system in the making but not an independent cluster of stars similar to our own 'grindstone'. He also found that even with the best of his instruments he could not resolve all the nebulae into stars or find stars in them. Some must be clouds of gas.

Herschel gave up on his grindstone model later, when he discovered that many double stars are true binaries (two stars orbiting the same centre of mass) with the pair of stars clearly the same distance from Earth yet differing in brightness. He was forced to conclude that stars do not all have the same absolute magnitude. At the same time, his larger telescope was revealing that beyond what he had previously thought were the furthest limits of the universe the stars go on and on. He could find no end to them. Such was Herschel's influence that other astronomers also abandoned his grindstone model. The question of structure on this scale didn't resurface in a significant way again until the middle of the 19th century, when another great amateur revived it.

William Parsons, the third Earl of Rosse, who resided at Birr Castle in Ireland, was the feudal lord of the village of Parsonstown. He had been educated at Dublin and Oxford and served as a Member of Parliament while still an undergraduate. In 1841, at the age of 41, he succeeded to the earldom, which gave him free time and an ample independent income to pursue his passion for astronomy. Lord Rosse also had a good knowledge of engineering and plenty of space to build a foundry and workshops. The lack of a skilled workforce in Parsonstown he soon remedied by training the labourers on his estate. He did not intend to buy a telescope and install it at Birr Castle. He was going to design the thing, build it, and cast the mirror himself.

Though improvements in refracting telescopes had by then made them temporarily far more popular than reflectors for both observatories and private use, Lord Rosse's intention was to build a reflector larger than anything Herschel had used. Did he consider the weather in Ireland and wonder whether this was the most desirable location for the world's largest telescope? Later he wrote to his wife: 'The weather here is still vexatious: but not absolutely repulsive.'

Lord Rosse began experimenting with the construction of smaller telescopes and worked his way up. Eventually he achieved his goal – a tube 56 feet long and eight feet in diameter, with a six-foot mirror weighing four tons, set up to protrude like a giant cannon aimed at the sky from an amazing castle-like structure. This time there is no record of a concert. Though professionals in the field of astronomy scoffed that Lord Rosse was more interested in designing and constructing telescopes than using them, he began observing with his 'Leviathan of Parsontown' in 1845, even before the supporting structure was completed. He aimed his celestial cannon at the nebulae.

Lord Rosse knew that these faint, fuzzy patches in the sky had both intrigued and frustrated William Herschel. Herschel, his sister Caroline, and his son John, a fine astronomer in his own right who spent some years at the Cape of Good Hope surveying far southern skies, had catalogued thousands of nebulae. Nevertheless at the time Lord Rosse looked at them in the mid-19th century, both with his smaller telescopes and with his 'Leviathan', the nebulae were still one of the great enigmas of astronomy. Controversy continued over whether some were conglomerations of gas, perhaps not far away, which might be the birthplace of new stars and planets, or whether they were instead incredibly enormous clusters of stars, too distant to be resolved by an earthly telescope.

Lord Rosse saw the nebulae as no one had before. They were not mere clouds. By 1848 he had resolved 50 of them into stars. Some had complex structure and, as Lord Rosse went on observing and drawing what he observed, more and more of them turned out to be spiral, lens-shaped formations. It became impossible not to suspect that the then out-of-date idea that these were star formations similar to our own, and extremely distant, might be correct after all. It also seemed likely that our own star system was, like them, spiral and lens-shaped, which was remarkably close to the way Herschel had pictured it in his 'grindstone' model.

William Huggins, who like Lord Rosse had the wherewithal to build his own private observatory, and who had been one of the first to discover that light from the Sun and from other stars has similar spectral lines, also turned his attention to the nebulae. He analysed the light coming from the Orion nebula, the Crab nebula and others similar to them and found their spectra were like the spectra of hot, luminous gases, not the same sort of spectrum as light coming from the Sun and the stars. But he also discovered that light from other nebulae, the great Andromeda nebula for one, gave a continuous spectrum of the sort one would expect if it were made up of stars.

In 1885, the distant heavens gave earthly astronomers a spectacular opportunity, or at least many of them thought that was what it was. They had already judged that the Andromeda nebula, one of the largest of the spirals, was probably the closest. In this nebula a new star suddenly appeared and became bright enough to be just visible to the naked eye. Astronomers knew of only one kind of exploding star – a nova. Comparison of this star's brightness with the brightness of previous novae, and later with a nova in 1901, indicated that the Andromeda nova was relatively close to us. That meant of course that the whole Andromeda nebula was close, by some estimates the nearest thing outside the solar system – certainly not a distant formation as large as the Milky Way. All of which added to the confusion of what seemed to be conflicting spectral analyses.

While this study and speculation was going on, in the last quarter of the century, astronomers were beginning to realize the potential value of a fabulous new tool – photography. Back in 1839, Daguerre in France and Fox-Talbot in England had almost simultaneously announced their discoveries of the photographic process. That same year, William Herschel's son John Herschel took one of the earliest photographs, a view of his father's 40-foot telescope through the window of his house at Slough. (See illustration 6 in the plate section.)

Though there were some fine astronomical photographs in the mid-19th century, exposure times were not yet fast enough for photography to be of practical, routine use to astronomers. John Herschel's exposure time for the photo of the telescope was two hours. When Lord Rosse recorded his observations he did it with drawings, not photographs. But when in the 1870s the use of dry gelatin plates reduced the exposure time required in terrestrial photography to about $1/15$ second, a new epoch in astronomy began. It was no longer necessary to rely on words or drawings to share observations, or on memory to compare what a portion of sky had looked like on one night with its appearance on another. Photographs taken on successive nights or over a span of days, weeks and years allowed astronomers to study how the sky changed. Photographic records took the place of such descriptions as Galileo's of Jupiter's moons, or John Herschel's of the star Alpha Hydrae, in 1838, as he was sailing back to England from the Cape of Good Hope:

21 March Alpha Hydrae inferior to Delta Canis Majoris, brighter than Delta Argus and Gamma Leonis.

7 May Alpha Hydrae fainter than Beta Aurigae, very obviously fainter than Gamma Leonis, Polaris or Beta Ursae Minoris.

10 May Alpha Hydrae much inferior to Gamma Leonis, rather inferior to Beta Aurigae. It is still about its minimum.

11 May Alpha Hydrae brighter than Beta Aurigae no doubt.

12 May Castor and Alpha Hydrae nearly equal.

'Very obviously fainter' . . . 'rather inferior' . . . 'much inferior' . . . 'brighter, no doubt' . . . 'nearly equal'. What a difference photography was to make in the precision with which observations could be reported and compared!

The English astronomer Isaac Roberts pioneered long-exposure photographs, allowing more light to enter the camera through the telescope than would happen in a quicker exposure. It was he, in 1888, who took the first photographs of Andromeda. Roberts' photographs confirmed that Andromeda is spiral in shape and clearly revealed the spiral arms in the galaxy's outer regions. However, even those photographs couldn't settle the question of what Andromeda actually *is*. Eleven years later, photography was first put to the task of recording a spectrogram of Andromeda, which indicated that it was a 'cluster of sun-like stars'. Yet Huggins had just previously seen mixed dark and light bands from Andromeda!

The confusion about the nebulae was part of a continuing debate about the larger picture. Friedrich von Struve, who had first measured the parallax of Vega, believed that the Milky Way's disc's edges extended to infinity, with interstellar matter absorbing the light from remote regions so that they remain eternally hidden from us. Others argued the pros and cons of a proposal that the Milky Way consisted of concentric rings of stars.

Partly because of the advent of photography, increasing attention was given not only to the position of stars but to their motions. It turned out that this motion was not, as previously thought, random. Stars were more likely to move in the plane of the Milky Way. In 1904, J.C. Kapteyn discovered that the majority of those stars which are easiest to observe move in two streams towards different parts of the sky. Like William Herschel he counted stars and found that Herschel had not been far off in his conclusions about their distribution.

Kapteyn thought the Sun was near the centre of the galaxy; while American astronomer Harlow Shapley would soon argue, based on study of globular clusters, that it was not. In 1913, the Dutch astronomer C. Easton concluded that the whole universe was one large spiral, shaped like the spiral nebulae. He thought that these nebulae were only miniatures of the greater spiral, within its boundaries.

The definitive answer was still almost quarter of a century away. However, at the turn of the twentieth century, astronomy was not far from a tremendous breakthrough when it came to establishing a foothold beyond the range of parallax measurement.

Two ingredients were necessary for any significant advance: first, the ability to identify a class or family of stars by some characteristic other than brightness – some characteristic certain not to change with distance; second, the ability to measure the distance to at least a few of the stars in that family or, failing that, at least to decide that a number of stars in the family were all approximately the same distance from us.

In the closing years of the 19th century, astronomers were having some success on all of these fronts. They were using the parallax method to measure the distances to as many stars as possible within parallax range and making catalogues of these stars and distances. They were continuing the search for distinguishing characteristics that could be depended on not to change with distance – colour perhaps, or some pattern of variation of colour or brightness, or patterns of spectral lines. And they were attempting to identify groupings of stars that are all approximately the same distance from us. With this combination of efforts, researchers were managing to edge themselves further and further into the cosmos.

There was a risk, just as there was in the analogy with the elephants and the giraffes. Discovering a weakness in one rung of the cosmic distance ladder could, and would several times, necessitate recalibrating the entire structure. But this was a problem astronomers had learned to live with, making repeated adjustments, hoping their margins of error were no greater than they estimated and that somewhere along the line there would be independent evidence to show that their measurements hadn't been far from the mark. Things were going fairly well. They were about to get much better.

In the early-to-mid-19th century, the fashion for observatory building had spread to the United States. At first most of the telescopes there were imported from Europe. In the 1830s there were good refracting telescopes at Yale University and at Wesleyan University in Middletown, Connecticut, but no actual 'observatories' at either place. Yale just stuck its telescope out of a window. In 1838, Williams College, in Williamstown, Massachusetts, opened its Hopkins Observatory, housing a 10-foot Herschel reflector bought by Professor Albert Hopkins in England. The Harvard College Observatory was founded in 1839 but didn't have a building or very much in the way of equipment. Earlier, in 1815, a delegation had gone to England to purchase a telescope for Harvard, found the desirable instruments too expensive, and went back empty-handed. In 1843 the local citizens of Cambridge, Massachusetts, disgruntled that there was no telescope around through which they could view the Great Comet of that year, offered to share with the university the cost of purchasing one. Harvard accepted, and the telescope was acquired from a distinguished firm in Germany.

By the 1890s the Harvard College Observatory had become a world-class institution. It was there that new rungs on the cosmic distance ladder were about to be nailed in place.

Henrietta Swan Leavitt was born in 1868 and studied at what would later become Radcliffe College in Boston, then known as the Society for the Collegiate Instruction of Women. At the nearby Harvard College Observatory, the eminent astronomer Edward Pickering was cataloguing and analysing stars and mentoring younger scholars.

There were few if any women among these budding astronomers, though women were hired to do the painstaking drudgery of writing down in endless rows of figures the positions and brightnesses of stars. However, women employed by Edward Pickering sometimes had a chance to do more creative work, for occasionally he encouraged someone from among his volunteer or underpaid female clerical staff to take on a more challenging assignment. In 1895 Henrietta Leavitt became a member of Pickering's staff. She started as a volunteer, received a permanent paid position in 1902, and soon became head of a department.

In 1908, Leavitt was looking for stars that varied in brightness, hoping to find a group of them that were all approximately the same distance away. It was logical to assume that all the stars in one of the Magellanic Clouds were, by cosmic standards, approximately the same distance away.

The Magellanic Clouds are two star formations that are not visible at any time of year in most of the northern hemisphere, where they never rise above the horizon. Seen from the southern hemisphere, they are large, misty smudges of light that could be mistaken for thin veil-like clouds faintly lit by the Moon on a fair night. The Australian aborigines believed that the Large Cloud was a part of the Milky Way that had been torn away. Europeans knew of the existence of these clouds before Ferdinand Magellan's voyage around South America in 1521 and called them the Cape Clouds. But Magellan's official recorder Antonio Pigafetta suggested that they be renamed the Clouds of Magellan in honour of that great explorer, who died just short of completing his circumnavigation of the globe.

Astronomers didn't pay the clouds much attention until William Herschel's son John studied the southern hemisphere skies in the 1830s and hypothesized that these clouds were fragments detached from the Milky Way. This was not plagiarism from the Australian aborigines. Herschel was in South Africa, not Australia. The younger Herschel thought this might mean the Milky Way was breaking up and that his father had been right to speculate that it couldn't last forever . . . indeed, that the past might not be infinite either.

By the end of the century, it was generally agreed that the Magellanic Clouds were made up of stars. There was less consensus about whether they were part of the Milky Way system or distinct from it but closely related. It was Edward Pickering's cataloguing, plus some of the techniques for measuring distances to groups of stars, that began to provide a clearer understanding of the Magellanic Clouds as the new century began. Today astronomers measure their distance from us at about 169,000 light years and consider them satellite galaxies of the Milky Way, but the earlier name has stuck – the Large and Small Magellanic *Clouds*. When Leavitt examined them, their distance was still unknown.

The reasoning behind Henrietta Leavitt's study was that if stars in the Magellanic Clouds were all approximately the same distance from Earth, then it was not differences in distance that caused some of them to look brighter than others. It seemed safe to conclude that stars that looked bright there really did have greater absolute magnitude than stars that looked dim there, and meaningful comparisons could be made between their apparent magnitudes. The Magellanic Clouds were close enough for individual stars to be identified and studied, though not close enough for their distances to be measured by direct parallax.

In the closing years of the 19th century and the early years of the 20th, the Harvard College Observatory had an outpost known as its Southern Station in Arequipa, Peru. By then photography had come into wide use in astronomy, not only allowing researchers to compare observations much more systematically than before but also making it possible for important discoveries to be made away from the telescope itself. From plates taken in Arequipa, Leavitt, in Boston, was able to identify 2,400 variable stars in the Small Magellanic Cloud.

Leavitt found that some of these variables had a remarkably dependable pattern to their variation: a steep increase to maximum brightness, and then a more gradual fall-off in brightness. Some took much longer than others to complete the pattern, and their range of brightness was also different. Nevertheless, the overall pattern was recognizable enough to set them apart. Leavitt noticed that among her sample in the Small Magellanic Cloud, the brighter a star of this type was, the longer it took to complete the pattern. The brightest ones took almost a month (some are now known to take over three months), the faintest only a day or so.

This relationship between a star's period (the length of time it takes a star to complete the cycle) and its brightness was the sort of clue for which Leavitt and others had been searching. The period of pulsation was a characteristic that wouldn't change with distance. Leavitt found 25 stars of this distinctive family in the Small Magellanic Cloud, all of which showed a clear relationship between brightness and period. She compared these 'light curves' with those of previously discovered variable stars nearer to Earth, in the Milky Way Galaxy, and found a match in the star Delta Cephei. 'Cepheids' is hence the name given to these variable stars.

Leavitt published her initial findings in 1908. Four years later, in 1912, she had compiled enough evidence to show that Cepheids could potentially provide a much more reliable way than any known before to pin down distances both in the Galaxy and far beyond it.

Wasn't it working backwards to have a discovery about stars so far away lead to a method to measure distances closer to home? One reason things happened this way is that studying the stars in the Galaxy and comparing their distances can be an extremely confusing undertaking. They are certainly not all the same distance from Earth, and brightness is no gauge of their distance. Take two hypothetical stars, Star A and Star B. Let's say that Star A's absolute magnitude is twice as great as Star B's — Star A is a *much* brighter star. Nevertheless, if Star A is twice as far away from us as Star B, Star A will actually look the fainter of the two to us on the Earth. (Recall the inverse square law and the discussion of the two lightbulbs.) The key to Leavitt's discovery was finding a family of stars with a good sample of its members in an area where she knew all the stars were approximately the same distance from us. The Magellanic Clouds were such an area, and there was no area like that within the Milky Way.

It might seem, however, that the Magellanic Clouds are surely large enough so that differences in stars' apparent magnitudes might deceive us there, just as they do in the Milky Way Galaxy. However, referring to our hypothetical situation above, *no* star in a Magellanic Cloud is twice as far away from us as another in the same Cloud. Think of it this way: if I look out of my window here in the eastern United States, I can say that the fence is twice as far away as the garage. Looking at them from Tokyo, I could not say that. For all intents and purposes, from Tokyo this fence and garage are the same distance away. So it is with the Magellanic Clouds. The stars there are so far away that we can treat them as being all the same distance from us.

Leavitt had found in the Small Magellanic Cloud that Cepheids with the same range of absolute magnitude had the same period of variation, which meant that if she knew how one Cepheid's period related to another Cepheid's period, she also knew the relationship between their absolute magnitude. For example, Leavitt found in her sample that if one Cepheid had a period of three days, and another had a period of 30 days, the second was six times brighter than the first. This meant that anywhere else in the sky she discovered a Cepheid variable star, she could measure its period of variation and be fairly sure that that would tell her how bright that star would appear *if* it were sitting among the stars in the Small Magellanic Cloud, and how its actual distance compared with stars there. This was definitely an advance in measuring stellar distances, but, again, as with Kepler's third law and Herschel's siriometers, what Leavitt had was a system of *relationships*, not absolute distance measurements. No one then knew the exact distance to the Magellanic Clouds, or the distance or absolute magnitude of any Cepheid in the Milky Way Galaxy. No Cepheid – not even Polaris, the North Star, the nearest Cepheid – was close enough to be measured by the parallax method. The cosmic distance ladder had been extended considerably, but it didn't reach the ground.

Not long after Leavitt's discovery, the Danish astronomer Ejnar Hertzsprung sought to remedy this situation. He applied a variation of the statistical parallax method and estimated the distance to two Cepheids. Using the relationships between period and absolute magnitude that Leavitt had established, he went on to calculate that the Small Magellanic Cloud is 30,000 light years away. That distance, though far short of the nearly 169,000 light years astronomers currently measure it to be, was much greater than anyone had been expecting.

Leavitt's 'standard candles', as she dubbed the Cepheid variables, almost immediately drew the attention of other astronomers, and one of them was Harlow Shapley. Shapley, born in Missouri in 1885, had come to astronomy by a serendipitous path. As a teenager he'd had a job as a crime reporter for a newspaper in Kansas, and he'd thought his formal education would be in journalism. Shapley arrived at the University of Missouri in 1907, but found that the school of journalism hadn't been built yet. When he returned a year later, nothing had changed, and Shapley was out of money and patience. As he later described it, he was 'all dressed up for a university education and nowhere to go'. He opened the university catalogue and began reading about the courses, starting at the front with the letter A. Archaeology didn't appeal. On the next page was Astronomy.

Four years later, with both a BA and an MA from Missouri, Shapley moved on to Princeton, and it was there that he began to investigate a different kind of variable from the Cepheids Leavitt had been studying. He concentrated on systems called 'eclipsing binaries' in which a pair of stars orbit one another (or, to be more exact, orbit their common centre of mass) with one star periodically eclipsing the other. The effect is that of a variable, for the light output changes as the stars eclipse one another. Shapley wanted to find out whether Cepheid variables might actually be binaries.

In 1914 Shapley's work came to the attention of George Ellery Hale, at the Mount Wilson Observatory in California. Hale had been the force behind the construction of the observatory, and at his invitation Shapley came to Mount Wilson to work with the 60-inch reflecting telescope there ('60-inch' refers to the size of the mirror), the largest telescope in the world at the time, for Lord Rosse's had fallen into disrepair. Soon Shapley was able to satisfy himself that Cepheid variables were not binaries.

Astronomers now know that Cepheids are elderly stars that have got past their 'main sequence' phase (the lengthy phase during which a star steadily converts its hydrogen to helium) and become 'red giants'. When the main sequence phase ends, the star starts to shrink. As it does, it heats up, and the heat flows into the outer layers of the star, energizing atoms of 'singly ionized' helium. 'Singly ionized' means there is an electron missing from these atoms. The increase in energy knocks yet another electron off, leaving the atoms 'doubly ionized'. Doubly ionized atoms readily absorb light. As a result of that absorption, the atmosphere of the star becomes opaque, holds in heat like a thermal blanket, gets hotter still, and expands. The star's outer layers swell, billowing out to about a hundred times the star's earlier size. As the star expands, it cools, for the energy has more room to spread out, and as the helium atoms cool they once again change from a doubly ionized state to a singly ionized state. The atmosphere becomes transparent again and begins to shrink, and the cycle starts all over again. That is the mechanism behind the pulsation of a Cepheid. It swells and shrinks and changes brightness repeatedly at a steady rate.

Shapley, like Hertzsprung, decided to try to measure actual distances to Cepheids, using a method of his own. Our Sun is not a Cepheid variable, but knowing something about the reasons for Cepheids' pulsation (though not the full explanation known today) gave Shapley a new way to compare a Cepheid with the Sun. Having the distance and size of the Sun, he found a way to calculate a Cepheid's absolute magnitude. Cepheids turned out to be some of the brightest of all stars.

With his new understanding and the finest telescope in the

world, Shapley began looking for a rich source of Cepheids and found it – the unattractively named 'globular clusters'. These are dense jewel-like spherical clusters of stars. Using the Cepheids in them as standard candles, Shapley began calculating their distances.

He made three discoveries about the globular clusters that he thought significant. First, in all the globular clusters to which he could calculate the distances, the brightest stars were always approximately the same absolute magnitude. Shapley had a new yardstick. Even if he didn't see a Cepheid in a cluster, he could assume that the absolute magnitude of the brightest stars there was the same as that for the brightest stars in clusters he had already been able to measure. Shapley's second discovery about the globular clusters was that they appeared to be spread equally above and below the plane of the Milky Way, and this seemed to indicate that they were part of the same system. Thirdly, some of them were distributed much further out than the stars of the Milky Way – so far out, in fact, that later in the century there would be confusion as to whether some group-ings of stars were globular clusters in the Galaxy's halo or separate dwarf galaxies.

It occurred to Shapley that globular clusters might form a skeletal outline to the Milky Way. Relative to the solar system, the distribution of the clusters appeared to be lopsided. A great many of them were gathered in an enormous sphere far from the solar system, a cluster of clusters centred on a point in the direction of the constellation Sagittarius. Shapley published his results in 1918 and 1919, interpreting them to mean that the centre of this spherical group of globular clusters was the centre of the Milky Way system. Our Sun was nowhere near that centre.

We hold our breath waiting for the clamour of opposition and controversy that surely must follow such an announcement. Humanity evicted once again – forced to move down yet another notch from our original exalted position in the centre of the universe! It's surprising to learn that there was little public or scholarly outcry about Shapley's discovery, though there were conflicting theories at the time.

What *did* shock astronomers and provoke a great deal of controversy was Shapley's measurement of the size of the Milky Way system. It was so huge that Shapley surmised that the system and its outline of globular clusters – in all, by his calculation, 300,000 light years across – must indeed be all there was to the whole universe. Nebulae such as Andromeda were not independent systems. At most they were minor satellites of the Milky Way. We now know that Shapley overestimated the size of the Galaxy because he failed to take into account the way intervening dust affects light coming to us from the globular clusters. The dust dims the light, and the sources looks fainter and further away than they really are. Astronomers now measure the Galaxy at approximately 100,000 light years, only a third as large as Shapley thought it was.

Opposition to Shapley's estimate came from one Heber Curtis of the Lick Observatory, whose road to astronomy had been almost as odd as Shapley's. Curtis had been a professor of classics and Latin, but when the college at which he taught merged with the University of the Pacific, he changed hats and became a professor of astronomy and mathematics.

Curtis wouldn't accept that Cepheids could be used as reliable standard candles. He insisted that the spiral nebulae were other systems far outside our own, that our system was much smaller than Shapley's estimate, and that the solar system was the centre of the Milky Way. Curtis suggested that the 1885 nova in Andromeda, whose apparent magnitude had been taken as indication that the Andromeda nebula was well within the Milky Way system, might actually have been of much greater absolute magnitude than the nova in 1901 to which it was compared – and hence much further away than most astronomers thought.

Shapley scoffed at this suggestion. If the Andromeda nebula was another star system like our own, not nearby at all, the 1885 nova there would have had to have had an absolute magnitude equal to a billion ordinary stars. It was absurd to imagine it could have been that bright, or that the universe could be large enough to contain a great many independent systems as big as ours. Furthermore, if nebulae like Andromeda were rotating at the rate some observations indicated, and were as far away as Curtis was claiming, they had to be rotating at a speed faster than the speed of light. That was impossible.

Curtis, undiscouraged, proceeded to search for other novae in Andromeda, so as to have more to compare with the one there in 1885 and the nova of 1901. He discovered several, all much dimmer than the nova of 1885, and that added weight to his argument that the 1885 nova was indeed exceptionally bright. He insisted that it was the dimmer and much more common novae, not the uncharacteristically bright one, that should be compared with novae elsewhere in order to calculate the distance to Andromeda. He decided that Andromeda was hundreds of thousands of light years away, far outside the Milky Way. Shapley still disagreed.

In 1920, the arguments between Shapley and Curtis culminated in a debate arranged by the National Academy of Sciences in Washington DC – attended by Albert Einstein, for one. The debate settled nothing. Hindsight shows that each man was right about some things and wrong about others. Together they could have put together a good picture of the universe. Most thought that Shapley had come off the worse in the debate, though astronomers were soon agreeing that his location of the centre of the Milky Way Galaxy was correct. Shapley left Mount Wilson not long after the debate to become the director of the Harvard College Observatory. Pickering, Henrietta Leavitt's mentor there, had died in 1919.

The first two decades of the 20th century ended without a

resolution to the question of how large our star system is and whether there is anything else besides it in the universe. The nebulae were still a puzzle. Even as late as the 1920s, there continued to be opposition to the 'island universe' theory that argued that some of the nebulae were other systems on the scale of the Milky Way. This opposition had some observational data weighing in on its side.

There was evidence that the nebulae were insignificant in size compared to the Milky Way – not its equals at all. Earlier indications that some nebulae had the spectra of stars (and must not therefore be made up only of gas) were called into question by the discovery that nebulae sometimes reflect light from elsewhere, and what is seen from them isn't all their own light. There were also Shapley's arguments: the enormous size of the Galaxy made it highly improbable that there should be others like it, and the speed of rotation in a spiral galaxy, were it sufficiently distant to put it outside the Milky Way, would be greater than light speed.

Waiting in part on the unravelling of these mysteries, there was an upheaval in the making that would rival Copernicus's *De revolutionibus* as a watershed in the history of astronomy.

Back near the beginning of the century, Vesto Melvin Slipher, a young Midwesterner with only an undergraduate degree from the University of Indiana, came to work at the Lowell Observatory in Flagstaff, Arizona. Slipher was a quiet man, methodical and meticulous. When it came to insisting on tying up loose ends before announcing a discovery, he rivalled Copernicus. Slipher spent his entire career at Lowell. While working there he earned his MA and PhD from Indiana. In 1916 he became the observatory's acting director; and in 1926, its director.

The wealthy astronomer Percival Lowell, who had built the observatory that bore his name and, at the time he hired Slipher, was its director, was one of those who believed the nebulae could be other solar systems in an earlier stage of formation. He set Slipher to work measuring spectra of spiral nebulae, looking for Doppler shifts in their light.

This was no easy assignment. Slipher couldn't merely point the telescope at a nebula and snap the camera. Exposures lasted for 20 to 40 hours, over several nights. A researcher couldn't stray far from the unheated telescope dome – purposely left cold because heat could mar the image – for he had to be certain that the nebula remained at the centre of the field of vision. The result at best was a faint, diffuse image of the nebula, not a concentrated point of light like a star. Using the spectroscope to spread it out further made it even fainter. The spectral lines were hard to identify, sometimes too faint if the spread-out image was too large. On the other hand, an image sufficiently bright for the lines to stand out clearly was apt to be too small for the shift in them to be measured.

In spite of these obstacles, in 1912 Slipher succeeded in obtaining four spectrograms showing the Doppler shift in the light from the Andromeda nebula. Readers who are anticipating what is to come in this book may be brought up short by the news that all four showed the light to be blue-shifted – indicating that Andromeda is moving towards us.

To review briefly: what Slipher saw in the spectra of light coming from the Andromeda nebula were familiar patterns of absorption lines produced when light passes through particular gases, but the patterns were shifted towards the blue end of the spectrum. (Review Figures 4.6 and 4.7.) Doppler and Fizeau had shown that such a shift signifies movement towards us along our line of sight, and Slipher accordingly interpreted this particular shift to mean that the distance between Earth and the Andromeda nebula was growing smaller.

Between 1912 and 1914 Slipher painstakingly pushed his equipment to its limits and measured Doppler shifts of 12 more nebulae. Andromeda proved to be an exception. Only one other had a blue shift. The others were red-shifted. Slipher calculated that they were rushing away at speeds of hundreds of kilometres a second.

Slipher, characteristically, didn't jump to conclusions. Thirteen nebulae among all those known in the skies was too small a sample to claim that it indicated all or most nebulae were receding from Earth. However, in 1914, with characteristic modesty, he reported his findings to the American Astronomical Society. John Miller, who had been one of Slipher's professors, described the event: 'Something happened which I have never seen before or since at a scientific meeting. Everyone stood up and cheered.'

Slipher went on designing and improving his own instruments and continued to find that most of the nebulae he was able to study did indeed show red shifts. In early 1921 he reported a nebula that according to his calculations was increasing its distance at a speed of approximately 2,000 kilometres per second. In 1922, he sent Arthur Eddington, an eminent physicist at Cambridge, measurements for 40 spiral nebulae, 36 of which were receding. Eddington was intrigued. He also was a cautious man, but he went so far as to suggest that this discovery about the nebulae, which were widely thought to be the most remote objects yet known, might be a hint about 'general properties of the world', by which he meant 'universe'.

By 1925, astronomers had measured 45 nebular Doppler shifts – Slipher 41 of them, other astronomers the remaining four. The score was now 43 red shifts to two blue shifts. What might have been a coincidence was definitely beginning to look like a trend.

Clearly Slipher had made a discovery of enormous importance, but it wasn't obvious at first what it signified. Slipher's own initial interpretation was that the drift of the solar system through space was increasing the distance between it and the nebulae. One problem with interpreting the significance of the red shift was that knowing the nebulae were moving away from us, or we from them, still didn't tell us how far away they were or what they were.

Now it so happened that when Slipher first announced his findings about red shifts to the American Astronomical Society in 1914, a young man named Edwin Hubble was in the audience. Hubble's background was unusual. His earlier university training hadn't been in astronomy at all. By the time he came to astronomy full time he was already a highly successful lawyer.

Born in Missouri in 1889, Hubble attended both high school and university in Chicago and then went to Oxford on a Rhodes scholarship. He was a polymath who was ready to tackle just about anything, including tank diving and high-level amateur boxing. But it was astronomy that won him in the end. He practised law for only a few months before going back to the University of Chicago to study astronomy and work as a research assistant at the Yerkes Observatory. 'All I want is astronomy,' he said. 'I would much rather be a second-rate astronomer than a first-rate lawyer.' In 1917, with a PhD in hand and an offer of a job at Mount Wilson, Hubble first went off to France to fight in World War I, returning in 1919. He came to Mount Wilson just before Shapley left there to take up the post at Harvard.

With the 60-inch reflecting telescope at Mount Wilson, and sometimes the more powerful 100-inch reflector that had just come into service in 1918, Hubble began to investigate the nebulae. By 1922 he had confirmed that nebulae that do not have a spiral structure don't shine with their own light. Either they shine by light reflected from stars within or near them, or they absorb enough energy from nearby stars to cause the hot gas of which they are composed to glow. Hubble's study confirmed earlier strong suspicions that these nebulae are part of the Milky Way system, but that didn't settle the question of the spiral nebulae. Hubble turned his attention to those next.

He was sure that some of the spiral nebulae were made up of

stars, but for many spirals, even by using a magnifying glass to scrutinize the best photographs he was getting with the 100-inch telescope, he couldn't produce the sort of evidence that would make him, and others, certain this was so. Hubble decided to investigate a faint patch of light called NGC6822, which *could* be resolved into stars. And there, in 1923, he found some of those stars varying in brightness. At first he failed to recognize their significance. He turned instead to the Andromeda nebula, where other astronomers had been discovering faint novae.

In the autumn of 1923, Hubble was using the 100-inch telescope to photograph Andromeda night after night. His observations were part of a survey to search for novae there that might be used to test out Curtis's ideas about nebulae. Hubble almost immediately found a couple of novae and another faint object that he at first thought was a third. At this point he decided to delve into the Mount Wilson archives to look for this star on older photographic plates. The comparison showed that it was actually a variable star, a Cepheid with a period of approximately a month, which meant that its absolute magnitude at its brightest was about 7,000 times as bright as the Sun. In order for it to appear as faint as it did, it had to be about 900,000 light years away. Hubble looked again at the photographs he had recently taken of the nebula NGC6822, and this time he recognized the varying stars as Cepheids, allowing him to calculate that NGC6822 was about 700,000 light years away.

These measurements to Andromeda and NGC6822 would later prove to be underestimates. Nevertheless they settled the question whether the spiral nebulae are in the Galaxy or are remote independent 'island universes' – other galaxies. By Hubble's measurement the Andromeda nebula was much further away than any star in the Milky Way. That indistinct, oval blur that we see from the northern hemisphere was definitely another galaxy – a collection of millions of stars. Astronomers now measure its distance as about two and a quarter million light years. The Andromeda galaxy is the furthest object visible from Earth with the naked eye. At that distance, the nova of 1885 in Andromeda had to have been much brighter than any ordinary nova. It was in fact something rarer, a supernova, the explosion of a star at the end of its life, as bright as nearly a billion Suns.

Hubble rushed news of his discovery about Andromeda and NGC6822 to Harlow Shapley. On first reading Hubble's message, Shapley turned to his colleague Cecilia Payne-Gaposhkin and commented, 'Here is the letter that has destroyed my Universe.'

Hubble christened Andromeda and other similar independent systems 'extragalactic nebulae'. By the end of the year he had resolved the outer part of the Andromeda nebula into stars, and a year later he was confident enough about the nature of the spiral nebulae in general to present his findings to the American Astronomical Society. His paper won an award donated by the American Association for the Advancement of Science to the two most outstanding and important papers presented at the meeting. The other winner was a paper on the digestive tracts of termites.

Over the next five years Hubble went on accumulating evidence and began developing techniques for estimating the distances to galaxies out beyond the range in which individual stars could be seen and identified as Cepheids. One technique resembled the one Shapley had employed when estimating the distances to the globular clusters: assuming that the brightest stars in all galaxies were approximately the same absolute magnitude and using these stars as distance indicators. That allowed Hubble to measure galaxies four times as remote as the furthest galaxy in which he could find a Cepheid. He estimated this range as about 10 million light years.

Hubble didn't stop there. He began hammering new rungs into the cosmic distance ladder at a furious pace. He decided that globular clusters could be used as a standard, assuming that the brightest globular clusters in all galaxies are approximately the same absolute magnitude. To go further yet, Hubble decided to assume that all galaxies have approximately the same absolute magnitude, or at least all fall within a narrow range. Recognizing that there would be some amount of error in this method, he nevertheless calculated that his distances would be no more than three times too large or three times too small. Hubble estimated that his techniques took him out to about 500 million light years – a volume of space containing about 100 million galaxies. Others would later refine his method and measure even further by assuming that the brightest galaxy in a galaxy cluster has approximately the same absolute magnitude as the brightest galaxy in every other cluster.

The accuracy of all this measurement stood or fell on the reliability of the measurement of the distance to Cepheid variables, using a variation of the statistical parallax technique. The whole ladder stood on that footing. The Cepheid yardstick has been revised several times over the years. Nevertheless, however rough the initial measurements were, Hubble did establish once and for all that the universe extends for billions of light years. His results seemed to indicate that the distribution of galaxies and galaxy clusters is fairly uniform throughout space. Hubble's investigations and those of his successors into the nature of Andromeda and other 'extragalactic nebulae' also turned a mirror on our own Galaxy, helping astronomers get a better idea of what it must be like as a whole, for of course they couldn't see it from a distance in space as they could Andromeda.

In the years following Hubble's discovery that at least some of the nebulae are well outside the Milky Way, and the coinciding realization that our own system is one galaxy among many others, some astronomers were still reluctant to trust Cepheids as reliable distance calibrators. There was one particularly nagging problem. As more and more galaxies were measured, most of them turned out to be quite a bit smaller than ours. Andromeda was only one sixth the size of the Milky Way. Others were smaller still. Was our Galaxy really exceptionally large? Perhaps by far the largest? That seemed suspicious enough for astronomers to have some doubts about the measurements.

By the 1940s the bright lights of the rapidly growing city of Los Angeles were making Mount Wilson a less-than-ideal location for a telescope. However, in the middle of the decade, at the height of World War II, these lights were often blacked out because of the threat of bombing raids. Though the citizens of Los Angeles undoubtedly found this unpleasant, it was a stroke of luck for astronomer Walter Baade, who because he was a German national had not been allowed to join the war effort in military research, but instead was left in almost solitary splendour at Mount Wilson, with the 100-inch telescope virtually all to himself.

Baade had been born in Schröttinghausen, Germany, and got his PhD from the University of Göttingen in 1919. In 1931, with the political climate in Germany changing, he had come to the United States to work at the Mount Wilson and Palomar Observatories, where he spent 27 years before returning eventually to his native country.

Under ideal viewing conditions during the blackouts, Baade studied the stars in the Andromeda galaxy through the 100-inch telescope. He found that those in the centre of that galaxy and in what he called its 'outer skeleton' or halo tended to be red and yellow, while those in the spiral arms were white and intense blue. Baade concluded that the stars must belong to two different 'populations'. He christened the white and blue stars 'Population I' stars. These turn out to be hot young and middle-aged stars. The red and yellow ones he called 'Population II' stars. These are much older stars.

Beginning in 1948, Baade was one of the first to use the 200-inch reflecting telescope (known as the Hale Telescope) on Mount Palomar, not far from Mount Wilson. He was puzzled to find Cepheid variables, with a difference, in the Andromeda galaxy halo. They were four times fainter than the Cepheids in the rest of the galaxy, the Cepheids Hubble had used to make his measurement. Baade, putting this finding together with what he had earlier discovered about the two different star populations, concluded that each population must have its own kind of Cepheid variable.

All Hubble and his associates had known were Population I Cepheids, the ones in the spiral arms of the Andromeda galaxy. The result was that they had been comparing apples and oranges, for the Cepheids they'd used for comparison were actually fainter Population II Cepheids. For this reason, Hubble's measurements of the distance and size of the Andromeda galaxy had been too small. Baade's discovery doubled the size and age of the universe. This increase helped clear up an embarrassing conundrum, for Hubble's measurements had made the universe younger than geologists knew the Earth to be.

Baade's discovery also led to a far better understanding of spiral galaxies, including our own. As astronomers recalculated their measurements, they found that spirals are all much larger than Hubble had estimated, and far more remote. While our mental picture of the universe swelled in size and distance, our picture of our own Galaxy shrank accordingly. Clearly, Shapley's 300,000 light year measurement for the diameter of the Galaxy was an overestimate. The diameter was closer to 100,000 light years. The Milky Way was not 10 times larger than most spirals and six times larger than the Andromeda galaxy. It was fairly average.

In 1952, Baade's former PhD student Allan Sandage joined the full-time staff at the Hale Observatories. Sandage, born in Oxford, Ohio, had known by the age of 11 that he wanted to be an astronomer when he grew up. When he looked through a friend's telescope, as he remembers it, 'a firestorm took place in my brain'. He spent two years at Miami University in Ohio and then two in the Navy as an electronics specialist. When World War II ended, Sandage finished his undergraduate degree at the University of Illinois, then headed west and enrolled in 1948 as one of Cal Tech's (the California Institute of Technology) first PhD candidates in astronomy. While he was still working on his degree he began collecting data at Mount Wilson for Baade, who was his thesis adviser. Then Edwin Hubble, who was observing with the 200-inch telescope, suffered a heart attack and needed a graduate student assistant to help him continue his work. Sandage was summoned to Mount Palomar. After a short sojourn at Princeton as a post-doctoral student, he came back as a member of the staff.

In 1958, five years after Hubble's death, Sandage discovered that some of what Hubble had thought were bright stars in distant galaxies and had used as measuring rods were instead glowing nebulae lit by many stars. That discovery more than tripled the size of the universe and increased its estimated age to about 13 billion years.

Human beings have a long history of underestimating the distance to objects in the heavens and the size of the universe as a whole. Though there have been a few overestimates, the scientific and popular image of the universe through the years has had to be revised upwards, by alarming increments, again and again. We have no intuitive feel for a distance of 13 billion light years, the size Sandage settled on in the late 1950s. But was that still too small? Or had he actually taken things too far? Could it be we would never know the answer? Nearly everyone who read the newspapers had been aware since the early 1930s that the universe wasn't cooperating as much as one might wish in this measuring venture. In more than one sense, it wasn't holding still to have its measurements taken!

CHAPTER 6

The Demise of Constancy and Stability
1929–1992

On philosophical grounds too I cannot see any good reason for preferring the Big Bang idea. Indeed it seems to me in the philosophical sense to be a distinctly unsatisfactory notion, since it puts the basic assumption out of sight where it can never be challenged by a direct appeal to observation.

Fred Hoyle

In the late 1920s, a sea-change was about to take place in the way men and women envisioned the universe. All of Edwin Hubble's earlier contributions might seem enough for one lifetime, but when it came to the impact he was to have on the history of human thought, Hubble had only barely begun.

Hubble was continuing his observational work in collaboration with Milton Humason, whose background, like Hubble's, was not in science. Humason was a janitor and mule driver at Mount Wilson whose formal education had ended at the age of 14 when he'd come to summer camp there and decided he didn't want to leave. By 1919, Humason, now 28, was a night assistant, tending the telescopes, assisting the astronomers, and occasionally doing a little observing on his own. So obvious was his extraordinary innate skill with delicate instruments and the large telescopes that George Ellery Hale, ignoring the opposition of those who thought someone of Humason's background should not be part of the academic staff, decided to appoint him Assistant Astronomer. In 1928 Humason began working with Hubble on the measurement of red shifts of faint distant galaxies.

In 1929, having established beyond doubt that there are many galaxies besides our own, the two men made the first announcement since Copernicus's *De revolutionibus* that rivalled that book's significance in the history of astronomy. Hubble and Humason had found that except for galaxies clustered close to ours every galaxy in the universe appears to be receding from us. What's more, on the large scale, every galaxy appears to be receding from every other. These discoveries had a much more immediate impact on the scientific community and the wider public than Copernicus's book, precipitating rapid change in ideas about what the universe is like, about its history, and even about ourselves.

It hadn't escaped Hubble's notice that there were connections between the observations that Slipher, Humason and he were making and the solutions that physicists such as Willem de Sitter, Alexander Friedmann and Abbé Georges Henri Lemaître were getting from the equations of Albert Einstein – solutions that implied that the universe must be either expanding or contracting.

The debate about whether the universe is expanding, shrinking or just holding its own has a history that goes back long before the 20th century. Ancient and medieval thinkers regarded the Earth as the region of the universe where change, decay and evil held sway, while all beyond the Moon was unchanging and perfect. Newton echoed these sentiments to the extent of believing that a universe created by God could not be changing dramatically over time, for constancy and stability reflected the nature of God, while change (implying decay and conflict) did not. Newton was also of the opinion that if a system goes far awry, as he realized the planetary orbits would over time, God would set it right again, so things would never be allowed to change too drastically.

As for the specific question whether the universe might be expanding or contracting, Newton decided on logical grounds that it couldn't be doing either. He reasoned that if it were expanding or contracting, there would have to be a centre to the motion – in other words, a point away from which it was expanding or towards which it was contracting. But matter distributed uniformly through an *infinite* space (as Newton believed it was) has no centre. Newton couldn't foresee that others nearly three centuries later would find that his own equations led to the prediction that the universe must be expanding or contracting. In the 18th century, Kant, who took off from Thomas Wright's picture of the universe as a flattened slab of stars, thought that if the universe were not perfectly balanced between the orbital motion of stars and their gravitational attraction for each other, it would end in destruction and chaos and lack 'the character of that stability which is the mark of the choice of God'.

Although it was out of fashion from the mid-19th century onwards to include God in scientific statements, the feeling that there was something sublimely rational and sacred about an unchanging universe and something shifty and distasteful about one that changed had by no means disappeared. It had become a doctrine of science rather than of religion. In an interesting turnabout, one of the 20th-century reasons for clinging doggedly to the notion of a static universe (one that isn't expanding or contracting) was that an expanding universe – which almost surely must have had a beginning – seemed more likely to require a creator. That was a possibility some had considered safely put to rest.

Albert Einstein resisted the idea of an expanding or

contracting universe for reasons having to do with his scientific intuition. Soon after he produced his general theory of relativity in 1915, Einstein and the Dutch astronomer Willem de Sitter realized that solutions to Einstein's equations implied that the universe was either expanding or contracting. Einstein was not necessarily one to cling to old assumptions, but at this juncture he did, and he dug in his heels. Annoyed by the ridiculous upshot of his equations, he wrote, 'To admit such a possibility seems senseless.' Such strong aversion did he feel that he decided to adjust his theory to cancel out the offensive prediction. He put in a new constant of nature – a 'cosmological constant', a mathematical term that would allow the universe to be static. He was later to regret this move, calling it 'the biggest blunder of my life'. But the notion of a cosmological constant didn't disappear when Einstein reneged on it. It still haunts physics.

While Einstein was tinkering with his equations, Russian mathematician Alexander Friedmann decided instead to take Einstein's theory at face value. Friedmann insisted that if there is a cosmological constant, its value is probably nothing else but zero. He pointed out in one of his first papers dealing with Einstein's theories that the assumption that the universe is static had always been only an assumption. No observations required one to believe it. Einstein himself was well aware that this was the case.

Friedmann proceeded to find not one, but a number of solutions to the cosmological equations of general relativity. Each solution described a different sort of universe. See Figure 6.1.

Friedmann predicted that regardless of where we were to situate ourselves in the universe, in any galaxy, we would find the other galaxies receding from us. The further away a galaxy is from us, the faster it's receding, twice as far away, twice as fast. For an analogy, imagine a loaf of raisin bread rising in the oven. Sitting on any raisin while the dough rises and expands between

Figure 6.1

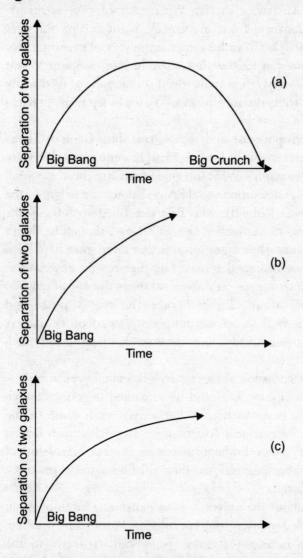

Three models of the universe: (a) The universe expands to a maximum size and then recollapses; (b) The universe expands rapidly and never stops expanding; (c) The universe expands at exactly a critical rate to avoid recollapse.

the raisins, we would see every other raisin moving away from us, twice as far, twice as fast. We cannot of course directly observe the universe from any vantage point except our solar system, but at least from here that is the sort of recession that Hubble observed in 1929 with the 100-inch telescope at Mount Wilson. The outward-bound speed of a galaxy is directly proportional to its distance from us. Twice as far away, twice as fast.

Belgian astrophysicist and theologian Abbé Georges Henri Lemaître discovered solutions to Einstein's equations that were similar to Friedmann's. What intrigued Lemaître most was what the equations and solutions could reveal about the origin of the universe. It was Lemaître who first described something like what was soon to be dubbed the 'Big Bang', though he didn't give it that name. His suggestion was that there must have been a time when everything that makes up the present universe was compressed into a space only about 30 times the size of our Sun – a 'primeval atom'. Partly because he was a priest and theologian as well as an astrophysicist, some of Lemaître's colleagues greeted his idea with derision. It smacked too much of Genesis.

Though Friedmann's theoretical work ended prematurely – he died at the age of 37 – and he remained largely unknown except among mathematicians, Lemaître's work came to the attention of observational astronomers, largely through British physics giant Arthur Eddington, whose student Lemaître had been at Cambridge, and another of Eddington's students, George McVittie.

Thinking about the universe as an expanding lump of raisin bread dough led to interesting speculation. Would it be possible, assuming the necessary technology existed, to travel to the surface of the loaf and find the border of the universe? What would be beyond that? Unfortunately those questions probably have no real meaning. Eddington fielded them by providing an analogy of a balloon and an ant:

The balloon has dots painted all over it. The ant crawls on the surface of the balloon. All that exists for this ant is that surface. It can't look outward from the balloon's surface or conceive of an interior to the balloon. Air is let into the balloon and the balloon expands. The ant sees every dot on the surface of the balloon moving away. Anywhere the ant crawls on the balloon, every dot is moving away. The ant may wander forever, like the Flying Dutchman, but it will never find an edge or a border to this universe. Our situation in our own universe is probably similar to the ant's, but with more dimensions. There is no edge from which we would see galaxies in one direction and absolutely nothing in the other.

The question of a 'centre' has cropped up often in this book, most recently in Newton's objections to an expanding or contracting universe. Where in the universe *did* the expansion begin? From what centre point is everything retreating? The Big Bang was an explosion that sent everything flying outwards. Even granting that there are no absolute directions in the universe, it would seem that beings riding on a piece of debris from this explosion would have the right to assume there is an answer to the question: Where did the explosion take place in relation to where we are now?

Eddington's balloon analogy helps with that question as well: Can the ant ask where on the balloon's surface the expansion began? No. From our vantage point, watching the ant, we can see that that question would be meaningless. No dot on the balloon represents the 'centre' of the expansion. Newton failed to imagine a situation in which all points in the universe are moving away from all other points, with no 'centre' to the expansion, no direction towards which we can look and insist that it all began there.

Paradoxically, living in an expanding universe means that there is a direction in which we *can* peer and see something different, perhaps even see an 'edge'. That direction is the past. What's more, in any space direction we look, we look towards the origin of the universe, for *any* direction is towards the past. That is true not only when we gaze deep into space with telescopes. It's true even in the small area of the room in which I write this paragraph. What I see of the opposite wall is old news. Of course the delay with which the picture of that wall reaches my eyes is not worth considering because light, and thus any picture that comes into my eyes, travels extremely fast, 186,000 miles or 300,000 kilometres per second.

When speaking of cosmic distances – where measurement in light years is more meaningful than measurement in miles or kilometres – light speed *isn't* terribly fast, and the delay can't be ignored. As the history of cosmic measurement continued in the 20th century, measurement of distance in space was to become inextricably bound up with measurement of distance in time. One can no longer ask how far away something is without also implying the other question: How far in the past is it? Questions about what is meant by an 'edge' or 'outside the universe' become entangled with questions about what is meant by a 'beginning' or 'before the universe'.

In the 1930s, many astronomers and theoretical physicists were taking Hubble's observations as direct evidence that the universe is expanding, but resistance to the idea had not ended and it was not all from within scientific circles. As was the case with Newton's *Principia*, the public were aware of stupendous changes going on in science. Einstein's theories were popular-ized in many forms and his name became a household word. When a new discovery or theory is fundamental enough to impinge on everyone's concept of the universe and reality – not just a few specialized scientists – there tends to be a feeling that others besides scientists should have a say about what is True in this matter.

Nearly everyone who reads popular science books remem-bers that Einstein didn't like the idea of an expanding universe. Fewer are aware of the ugly opposition from some who held political power. In 1936, in the Soviet Union, Joseph Stalin began a purge of scientists whose scientific findings and conclusions were not politically correct. One of the forbidden ideas was that the universe was expanding.

In his book *Fireside Astronomy*, British astronomer Patrick Moore tells of the experiences of his friend Nikolai Kozyrev. Kozyrev was an astrophysicist at the Pulkovo Observatory near St Petersburg. In November 1936 he was arrested and physically assaulted. In May 1937 he came to trial. What his offence was was never clearly stated, but he was sent to prison. After two years Kozyrev ended up in a labour camp, and there a fellow prisoner reported him for holding scientific views about an expanding universe that were contrary to Soviet doctrine.

Kozyrev was resentenced to 10 years' imprisonment. When he appealed, the sentence was changed to death. There was no firing squad at the labour camp, and a second appeal got the sentence reduced again to 10 years. Gregory Shain, later the Director of the Crimean Observatory, rescued Kozyrev from this appalling situation. He managed to get him transferred to Moscow in 1945 and saw him set free in 1947. Kozyrev returned to his work in astrophysics, having lost 10 years. Other Soviet scientists were less fortunate. Many were executed. The persecution even extended to the scientists' families. Kozyrev's wife was imprisoned, though not for such a long period as her husband.

Elsewhere the opposition was less extreme, and it soon focused not so much on whether the universe was expanding as upon whether it had a beginning. It was here that discoveries in astronomy and physics theory trod most seriously on philosophical and religious sensibilities.

One interpretation of Galileo's trial sees it as a clear contest between the authority of religion and the authority of science. In the 20th century, those who had thought science had won that contest long ago were chagrined to find science seeming to uphold a religious point of view. Anyone for whom the idea of a

God was anathema now had to face the unthinkable: a beginning . . . a moment of choice about whether there would be a
universe . . . a creator.

Not that all who found the Big Bang philosophically disquieting were self-declared atheists. Many men and women, often
without giving much thought to whether this conflicted with
their religious beliefs, had put their trust in the power of
science to explain the world. Since the time of Newton it had
been a growing assumption both in and out of science that
scientific laws and explanations underlie everything that occurs,
even those things that remain most mysterious and hidden, and
that, given time, human minds ought to be able to discover
those laws and explanations. The Big Bang threatened that
cherished assumption.

A passage from Robert Jastrow's 1978 book *God and the
Astronomers* sums up the situation. Jastrow is himself an astronomer and an agnostic, but he chides his colleagues for their
reaction: 'the response of the scientific mind – supposedly a
very objective mind – when evidence uncovered by science
itself leads to a conflict with the articles of faith in our
profession'. He goes on to say:

> This is an exceedingly strange development, unexpected
> by all but the theologians. They have always accepted the
> word of the Bible: In the beginning God created heaven
> and earth. To which St Augustine added, 'who can under-
> stand this mystery or explain it to others?' The develop-
> ment is unexpected because science has had such
> extraordinary success in tracing the chain of cause and
> effect backward in time . . . Now we would like to
> pursue that inquiry farther back in time, but the barrier to
> further progress seems insurmountable. It is not a matter
> of another year, another decade of work, another meas-
> urement, or another theory; at this moment it seems as
> though science will never be able to raise the curtain on

the mystery of creation. For the scientist who has lived by his faith in the power of reason, the story ends like a bad dream. He has scaled the mountains of ignorance, he is about to conquer the highest peak; as he pulls himself over the final rock, he is greeted by a band of theologians who have been sitting there for centuries.

Three Cambridge physicists declined the invitation to sit down with the theologians. Just as there had been excellent alternative ways of explaining Galileo's findings without having to have a moving Earth (Tycho Brahe's model, for example), there were excellent alternative ways of explaining Hubble's and Einstein's without having to have a beginning.

In 1948 Hermann Bondi and Thomas Gold, both originally from Austria, and Fred Hoyle introduced theories that allowed the expansion of the universe to happen without requiring that the universe have a beginning in time. Their 'Steady State' theory became the Big Bang's major competitor. According to Bondi, Gold and Hoyle's proposal, the universe hasn't always contained all the matter that is in it today. As the universe expands, new matter emerges to fill in the broadening gaps, and the average density of matter in the universe remains the same. While the stars in a galaxy like ours burn out and the galaxy dies, new galaxies are forming from the new matter. There would be no beginning or end to a Steady State universe. The unwelcome hint of 'creation' suggested by Big Bang theory would be eradicated.

For at least two decades, the scientific and philosophical debate went on between those who favoured one theory and those who insisted on the other, until finally, in the 1960s, new evidence came to light that Steady State theory could not explain and Big Bang theory actually had predicted. It should come as no surprise that Hoyle, one of the inventors of Steady State theory, was the author of one of the most insightful books about the Copernican revolution, a book with considerable sympathy for Ptolemy that points out clearly how *both* Copernicus and Ptolemy were correct.

The observational evidence that weighed in so heavily in favour of the Big Bang was not evidence from an optical telescope. By then astronomers had discovered that the old phrase 'I won't believe it until I see it' represented a ridiculously limiting attitude. Most of what goes on in the universe can't be 'seen' at all. It happens beyond the visible range of the spectrum.

It was no coincidence that studies of the heavens were for centuries only studies of visible light, and that radio astronomy was the first new astronomy to emerge. In only those two parts of the electromagnetic spectrum – the optical and radio ranges – are there wavelengths that can pass through the Earth's atmosphere. Radiation in the infrared range can reach as low as the highest mountains. The Earth's atmosphere blocks other radiation. Study of ultraviolet rays, X-rays and gamma rays coming from space can't be done at all without telescopes above the atmosphere. It wasn't possible to put them there until the late 1950s.

Radio astronomy began a quarter of a century before that, almost by accident. In the 1930s, trans-Atlantic phonecalls took place by radio transmission and were plagued by static. The task of finding out what was causing the static fell to Karl Jansky of the Bell Telephone Laboratories in Holmdel, New Jersey. Jansky built a special radio antenna – a long array of metal pipes – to aid him in his investigation.

As Jansky sorted out the static, he found that most of it came from thunderstorms, but there was also a faint hissing static that couldn't be so easily explained. The hiss was strongest when the region of the sky in the direction of the constellation Sagittarius was overhead. The central regions of our Galaxy lie in that direction. When this part of the sky was below the horizon, the hiss was weaker but it never disappeared entirely. The expectation prior to Jansky's discovery had been that the Sun would be the strongest source of radio waves in the sky, just as it is the brightest source of light. Now it seemed that a source could be very 'bright' in another part of the spectrum but show up not at all in terms of visible light. Jansky knew he had discovered the centre of the Galaxy. What else was out there that we were missing?

Though it might seem that such a mystery as Jansky's hiss would have made headlines, his work received little attention even within the astronomy community. The one exception to this apathy was Grote Reber, an eccentric bachelor and ham radio operator in Wheaton, Illinois, who read about Jansky's radio hiss in *Popular Astronomy* magazine. Reber proceeded to cash in his savings and design and construct his own device in his mother's back yard to listen to radio signals coming from the sky. Reber's telescope was a dish 30 feet (nine metres) in diameter. Astronomer Jesse Greenstein of Yerkes Observatory, when he later saw the back-yard telescope, called Reber 'the ideal American inventor. If he had not been interested in radio astronomy, he would have made a million dollars.'

In the 1940s, Greenstein tried to get Reber a position at the University of Chicago. That fell through when the University agreed to have Reber only if his pay and research support came from Washington, and Reber refused to go to the bother of explaining to bureaucrats precisely how the money for new telescopes would be used. Except for Reber, until the 1950s there was virtually no interest in radio astronomy in the United States.

Astronomers elsewhere in the world were quicker, after the development of radar during World War II, to appreciate the potential advantages of studying radiation in a part of the electromagnetic spectrum outside the visual range. In 1946 a radar signal was bounced off the Moon, and in the years that followed, the Jodrell Bank radio telescope in England led the world in mapping radio sources in space. Following close behind were Cambridge University and a team in Australia.

Particularly intense sources of radiation in the radio range of the spectrum became known as 'radio stars' and 'radio galaxies'.

A problem in early radio astronomy – and one reason it drew little interest at first from optical astronomers – was that radio telescopes like Reber's couldn't measure a source's position in the sky accurately enough to find out which visible object was emitting the radio waves. In order to accomplish that, there needed to be a hundred-fold improvement in resolution, and that meant a telescope about a kilometre in diameter. Radio waves come in a large range of lengths, but they are all longer than waves in the visible part of the spectrum, and their length is the problem. A radio telescope doesn't give good resolution unless it is quite a bit larger than the length of the radio waves it's receiving. Studying shorter radio waves only helps a little. The much shorter waves in the *visible* part of the spectrum allow optical telescopes to achieve resolution with relative ease. Radio astronomers finally solved this problem in 1949 by using networks of small telescopes linked to a receiving station which combines the signals. Such a network or array is a 'radio interferometer'.

England and Australia continued to dominate the field. In 1950 Martin Ryle of Cambridge University and his co-workers discovered evidence of radio emissions from four nearby galaxies including Andromeda. Ryle was an advocate of Big Bang theory, and as he and his colleagues continued to map the distribution of radio sources across the sky, their findings supported that theory. They discovered that radio galaxies are much more abundant in the far distance than they are nearby. Since the further we peer into space the further back in time we are looking, Ryle's radio telescopes were showing him the universe at a much earlier stage, and his discovery indicated that the density of radio galaxies was greater then than it is today. It's reasonable to think this should be the case if this is an expanding universe in which everything on the large scale is moving away from everything else and used to be much more closely crowded together.

In 1954, owing largely to the influence of Greenstein, by then a professor at the California Institute of Technology, the National Radio Astronomy Observatory was built in West Virginia and a radio interferometer was constructed near Yosemite National Park in California under the aegis of Cal Tech. But before that, even during the years when radio astronomy was virtually nonexistent in the United States, the Bell Telephone Company had continued to carry on research having to do with radio communications. It was at Bell Labs in New Jersey, where Jansky had first investigated the radio hiss from space in the 1930s, that the discovery took place that most scientists considered the clincher for the Big Bang theory.

Arno Penzias was born to a Jewish family in Munich. When he was six years old, he and his brother and parents were among the last Jews to get out of Nazi Germany, arriving as impoverished immigrants in the United States in the winter of 1940. Penzias received his undergraduate physics degree from the City College of New York and went on to Columbia University for his PhD. In 1961 he took a job at Bell Labs, where two years later he was joined by Texan Robert Wilson. Wilson had at first favoured the Steady State theory over the Big Bang. He was strongly impressed with Fred Hoyle, who had been a visiting professor at Cal Tech when Wilson was a graduate student there. In 1963, a year after completing his PhD, Wilson crossed the country to the east coast to work at Bell Labs.

At Bell there was a large, horn-shaped antenna designed for use with the Echo I communications satellite. In the spring of 1964, Penzias and Wilson were using the antenna to study noise levels that were hampering the satellite's transmission. Scientists working with the antenna had to make adjustments and limit themselves to signals that were stronger than the 'noise'. It was an annoyance that was possible to ignore, but Penzias and

Wilson chose not to. They noticed that the noise remained the same regardless of which direction they pointed the antenna. If the Earth's atmosphere itself were the source of the noise, an antenna pointed towards the horizon should pick up more noise, for it faces more of that atmosphere than an antenna pointed straight up. Penzias and Wilson concluded that the noise had to be coming either from beyond the Earth's atmosphere or from the antenna itself. Pigeons nesting in the antenna were evicted and their droppings cleared away. That didn't help.

Penzias and Wilson were unaware of theoretical predictions that had been made back in the late 1940s and also of current work going on in England, Russia and a few miles away from them in Princeton, New Jersey. In the 1940s, Russian-born physicist George Gamow, who had defected to the West in 1933, and Americans Ralph Alpher and Robert Herman had theorized about the early universe, running Friedmann's equations backwards towards the event with which the universe began. Their prediction was that there ought to be left-over radiation surviving from about 1,000 years after the origin of the universe. At that time, if Big Bang theory had it right, the universe was very hot, but by now the temperature of the radiation should have cooled to about five degrees above absolute zero. The prediction wasn't tested, for such radiation would not be easy to observe.

The idea that this radiation might indeed exist, and questions about its temperature, were still around in the early 1960s. In 1964, while Wilson and Penzias were tidying the pigeon droppings, Fred Hoyle and a colleague, Roger Taylor, in England were attempting to calculate what the background temperature of the universe would be today if it began in a Big Bang. And in the Soviet Union Yakov Borisovich Zel'dovich had concluded that given the abundances of hydrogen, helium and deuterium observed today, the universe must have begun in a hot Big Bang and its background temperature currently must be a few degrees above absolute zero. Soviet researchers had written papers about current radio astronomy measurements and what they implied about the background radiation, and they had even suggested that the most likely antenna in the world to be able to detect this radiation was the Bell Labs antenna in Holmdel.

Robert Dicke at Princeton was working along the same lines. Born in Missouri in 1916, he was of Hoyle's generation. He was educated at Princeton and at Rochester University, then worked on radar at the Massachusetts Institute of Technology (MIT) during World War II and after that joined the Princeton faculty. On and off through the years Dicke had considered the problems of the background temperature of the universe and the as yet undetected background radiation. In 1964, he set P.J.E. Peebles, a young researcher at Princeton, to work figuring out how the temperature might have changed over time in an expanding universe that had originated with a hot Big Bang. When Peebles had finished his calculations, Dicke gave two other researchers the job of setting up an antenna on the roof of the Princeton physics lab to try to detect the predicted radiation. It was at this juncture that Dicke received a phone call from Arno Penzias and Robert Wilson.

It so happened that Bernard Burke, another radio astronomer, had heard about Penzias and Wilson's puzzle, and he also knew (as they did not) of the work being done by Dicke. Burke proceeded to bring Dicke, Penzias and Wilson together, and they soon concluded that Penzias and Wilson had found by accident the radiation that Dicke had been hoping to discover.

This all-pervasive hum of radiation, coming at equal intensity from all over the sky, has been likened to a faded photograph of the universe as it existed about 300,000 years after the Big Bang. It is the oldest 'photograph' we have, and the most direct evidence that the universe was once much hotter and denser than it is now. The radiation has cooled with the expansion of the universe and been red-shifted so greatly that it reaches us in the microwave range of the spectrum at a temperature of about three degrees above absolute zero, a little cooler than the five degrees Gamow, Alpher and Herman had predicted in the 1940s.

What is the source of the radiation? According to Big Bang theory, the universe in its early stages was everywhere filled with electromagnetic radiation. This radiation wasn't in the visible part of the spectrum. It was far too hot for that, in the trillions of degrees. As space expanded, stretching the wavelengths of the radiation, it shifted through the spectrum. The universe gradually cooled, but until it was about 300,000 years old the radiation was still too energetic to allow electrons and protons to bind together and form atoms. If an electron began to orbit a proton it was knocked out of orbit by a photon – a particle of electromagnetic radiation. At about the 300,000-year mark, everything had cooled off enough so that photons no longer had the energy to knock electrons away from protons. Electrons and protons could form hydrogen nuclei and atoms, and photons could move about more freely. Physicists call this the 'decoupling' of radiation and matter. With that change, radiation (photons) streamed in all directions, and it is that radiation, red-shifted all the way across the spectrum to the microwave range, that we detect today as the cosmic microwave background radiation. Though that radiation began its journey longer ago and further away than anything else we observe, you don't need special equipment to detect it. The snow on a TV screen that appears when a station isn't broadcasting consists in part of this radiation, this dim afterglow of the Big Bang cataclysm.

It's an interesting bit of astronomy trivia that Wilson and Penzias were not actually the first to detect and measure the cosmic microwave background radiation. In 1961, another engineer at Bell Labs, Ed Ohm, had also tried to figure out what was causing the 'noise'. Eliminating everything that could be explained away, he found that it was the equivalent of radiation at a temperature of about three degrees above absolute zero. Unfortunately for Ohm, he didn't find the problem as annoying as Wilson and Penzias did, and he didn't pursue it or suspect its significance. No one put him in touch with anyone who knew of the theoretical predictions. It was Wilson and Penzias who shared a Nobel Prize for the discovery.

Why wasn't this discovery made sooner? Gamow, Alpher, Herman or Dicke probably could have discovered the cosmic microwave background radiation earlier had they tried. As it was, it took a combination of a private corporation – American Telephone and Telegraph (AT&T) – far-sighted enough to fund less practical science and attract such researchers as Wilson and Penzias, the stubborn curiosity of these two men themselves – who unlike Ohm weren't satisfied until they got to the bottom of a mystery – and a serendipitous meeting of observation and theoretical understanding when those *almost* were ships passing in the night.

From the early 1960s, everything seemed to fall into place for those who favoured Big Bang theory. The discovery that quasars – which theorists think are probably an early stage of galaxy formation – exist only at enormous distances from Earth was more support. According to Steady State theory, galaxies are periodically dying and being replaced by new galaxies made from new matter. If that were so, and if quasars are part of the process of galaxy formation, they ought to be fairly evenly distributed near and far throughout the universe. The fact that they are not argues against the Steady State and in favour of the Big Bang. Quasars' distance from us in space (and, by virtue of that fact, in time) means they only existed when the universe was much younger than it is now, indicating that this particular stage of galaxy formation occurred only in the distant past, hasn't happened again in later periods of the universe's history, and isn't still going on today. The universe is not repeating itself.

Astronomers and physicists have also continued to study the cosmic microwave background radiation for clues about how the universe has evolved. In 1973, Paul Richards and colleagues at Berkeley undertook balloon experiments to find out whether the spectrum of the background radiation was the spectrum that Big Bang theory predicted. They found that it was.

Even more support for the theory, also in the early 1970s, came from studies of the spectra of other galaxies to measure the abundances of various elements in them. Big Bang theory had predicted that about 25 per cent of the mass of all the elements making up the universe should be helium 4. The studies showed that prediction was on target. So did measurements within the Galaxy. Predictions of abundances of other elements such as deuterium, helium 3 and lithium also turned out to be what the theory prescribed.

While it looked increasingly likely that Steady State theory would go the way of Tycho Brahe's valiant last-ditch efforts to save the Earth-centred universe, the Big Bang theory wasn't entirely problem-free either. Two stumbling blocks were the 'horizon problem' and the 'flatness problem'.

The 'horizon problem' stems from the observation that the cosmic microwave background radiation is very homogeneous, the same in all directions in areas too far separated for radiation ever to have passed from one to the other even at the earliest moments. The intensity of radiation is so close to identical in those remote areas that it seems they must have exchanged energy and come to equilibrium. The question is: how?

The 'flatness problem' is the problem of why the universe has not either long ago collapsed again to a Big Crunch or else experienced such runaway expansion that gravity wouldn't have been able to pull any matter together to form stars. Neither has happened or seems to be happening at the moment. Yet having a universe somehow poised between those possibilities is so unlikely as to boggle the imagination.

A revised history of the Big Bang universe called 'inflation theory' proposed to solve both those problems. The idea emerged in the late 1970s, when Alan Guth, then a young physicist at the Stanford Linear Accelerator, reached the conclusion that the universe might early on have undergone a period of stupendous growth before settling down to the expansion rate it has today. Guth knew immediately that he'd hit upon a good thing. 'SPECTACULAR REALIZATION' he wrote in his notebook, and drew two concentric boxes around the letters.

Guth proceeded to work out a process which, at a time less than 10^{-30} seconds after the Big Bang (that number as a fraction is 1 as the numerator and 1 followed by 30 zeros as the denominator), could have caused gravity to become an enormous repulsive force. Instead of pulling matter back and slowing the expansion of the universe, it would, during a period lasting only an unimaginably small fraction of a second, have accelerated the expansion, causing violent, runaway inflation in the dimensions of the universe from a size smaller than a proton in the nucleus of an atom to about the size of a golf ball. When the inflationary period ended, the universe would continue to expand, but in the more sedate, familiar fashion.

Inflation theory helps the horizon problem by allowing the visible universe to have emerged from a region so tiny that it had an opportunity to reach equilibrium before it inflated. When it comes to the flatness problem, the theory says that out of an infinite number of possible universe stories – those that have the universe collapsing, those that have it expanding forever to thin oblivion, and the one in which it is perfectly poised between the two – the universe actually *can be expected* to be a universe perfectly balanced, expanding forever but at a continually decreasing rate, not collapsing and not eternally thinning out. Big Bang theory before inflation theory had not been able to help the universe walk that tightrope. But Guth and others who have contributed to the development of inflation theory explain that any imbalance between the expansive energy resulting from the Big Bang and the contracting force of gravity would have been wiped out by this period of runaway inflation, leaving the universe in that extremely unlikely and highly desirable condition of flatness. Highly desirable because that is the only sort of universe that eventually allows intelligent life to emerge.

To visualize the version of inflation theory that has the most to offer in this regard, first imagine the universe before the period of inflation begins, again using a balloon as an analogy. Inflate the balloon a little. That represents the expansion of the universe before the inflationary period. Pause to mark a tiny red dot on the surface of the balloon. Next attach the balloon to one of those machines that inflates balloons rapidly and turn the machine on at maximum force. That inflates the balloon to a truly remarkable size. The tiny red dot itself will have become huge. Imagine now that it isn't the whole balloon that represents all we observe and ever will observe of the universe. It is the red dot. Inflation theory asks us to believe that what we normally call 'the universe' is similarly only a tiny fraction of everything there is.

Let's suppose that instead of just one red dot we have drawn dots all over the balloon, perhaps an infinite number of them. Having blown up the balloon to enormous dimensions, will we find every dot representing something equal in size to the observable universe?

Andrei Linde of Stanford University has suggested that that may not be the case. Linde, a graduate of Moscow University and the P.N. Lebedev Physics Institute in Moscow, already had an enviable reputation in physics when he moved to Stanford in 1990. Among his many accomplishments — he also dabbles in sleight-of-hand magic, acrobatics and hypnosis — in 1983 he had introduced a new version of inflation theory.

His proposal was that the early universe before the period of inflation was in a chaotic condition, something like the surface of the ocean. If the universe was this chaotic, it would be ridiculous to talk about *the* initial state of the universe. We could find all sorts of initial states, depending upon which bit of it we examined. Local conditions for any of the dots on the balloon might be different from any other. The upshot is that when the gravitational repulsive force came, each dot would respond differently. Some not at all. But one version of the theory predicts that when the inflation ended, we would find that in any dot that *had* inflated, the force of gravity (now working in the more familiar way) and the repulsive force resulting from the original Big Bang explosion would be balanced in the way we now observe in our universe. Perhaps only one of all those dots would have been able to end up that way. If so, that dot is our universe.

Though inflation theory had great success explaining away problems in Big Bang theory, there was no observational evidence to assure anyone that inflation actually is *the correct* explanation. Discoveries in the late 1990s would throw theorists a new set of problems and perhaps new solutions, as well as some hope of observational evidence.

A third stubborn puzzle plaguing Big Bang theorists after Wilson and Penzias's discovery was how a universe that looked so uniform when it was 300,000 years old had become so diverse and clumpy all these years later. In repeated measurements, researchers found that the cosmic microwave background radiation was disappointingly uniform in temperature. The temperature was the same in readings taken out to the end of observability in every direction. This meant the early universe must have been extremely smooth, without lumps, clumps or irregularities that would show up as fluctuations in that temperature. How then could the universe have evolved to have galaxy clusters, galaxies, stars, planets – even such small clumps of matter as people? Somewhere back there must lie the seeds of those developments, but where?

Here is the problem: picture every particle of matter in the universe attracting every other by means of gravitational attraction. The closer to one another the particles are, the stronger they feel each other's gravitational pull. If all particles of matter in the universe are equidistant, and there are no areas in which a few particles have drawn together even slightly more densely, then every particle will feel equal pull from every direction and none will budge to move closer to any other particle.

It was this sort of gridlock researchers seemed to have discovered in the early universe, where matter appeared to have been distributed so evenly that it could never yield to form the structure evident in the universe today. If this were not so, why couldn't anyone find even the tiniest fluctuation in the background radiation – the 'photograph' of how matter was distributed back then?

Finally, in April 1992, astrophysicist George Smoot at Lawrence Berkeley Laboratory and the University of California at Berkeley announced that he and his cohorts at several other institutions had found the long-sought 'wrinkles'. New data from a satellite called the Cosmic Background Explorer (COBE) had revealed the fluctuations in the cosmic microwave background radiation that astrophysicists had been seeking for a quarter of a century. The temperature fluctuations measured no more than a hundred-thousandth of a degree, but that was enough, the researchers felt, to explain what had happened to the universe. These minuscule variations in its topography when it was only 300,000 years old were evidence of a gravitational situation in which matter could have attracted matter into larger and larger clumps.

How did the tiny fluctuations get there? You and I in our innocence might think that in any real-life situation, having a few irregularities is something to be expected. It would be much more surprising *not* to have them. Surely this is the sort of non-problem only a mathematical physicist who hasn't looked at the real world for a while would dream up! Even Isaac Newton commented that a situation in which matter is evenly spread is far less probable than getting a needle to stand on its point on a looking glass. But you and I, and Newton as well, for that matter, have been conditioned in a universe that is full of asymmetries and irregularities – the very situation modern theorists find suspicious. Why should the universe be like this?

There have been some stabs at answering the question of how the tiny fluctuations got there, but no observational evidence to help. Inflation theorists point out that according to the so-called Heisenberg uncertainty principle of quantum mechanics, what we call 'empty space' cannot actually be empty. Instead, always and everywhere in the universe there are tiny energy fluctuations. During the inflation period, the peaks and troughs generated by these fluctuations in the newborn universe would have been inflated large enough to serve as the seeds of all the irregularity to come.

In the year 2000, NASA is scheduled to launch its Microwave Anisotropy Probe. 'MAP' will measure the background radiation 30 times more precisely than the Cosmic Background Explorer did. Even more precise observations should come from the European Space Agency's Planck Satellite beginning in 2004. There are also several balloon-based missions in the offing.

Meanwhile, a wealth of evidence points to the fact that we do live in a Big Bang universe. To sum up its history, according to the theory as it stands amended by inflation theory: all that we observe or ever will be able to observe started out compressed in a state of almost unimaginable density. That exploded and everything – space itself – began to expand. After a short interval of extremely rapid inflation, the expansion slowed down and continued more sedately. Everything thinned out and cooled. All was smooth and virtually uniform except for some faint wrinkles, or 'density fluctuations'. While expansion continued, the gravitational attraction of areas where matter was already concentrated more densely pulled in more matter, and thus matter began clumping, eventually forming stars, galaxies, clusters and superclusters of galaxies gravitationally bound to one another.

While observations and experiments were confirming earlier Big Bang predictions, theorists had been raising new questions about the moment of 'beginning' itself. One in particular was: Does an expanding universe that is not a Steady State universe have to have had a beginning?

Both everyday logic and mathematics seemed to indicate that in a universe where on a large scale everything is moving further and further from everything else, if we could reverse the direction of time and travel back towards the beginning, we would find things getting closer and closer together. Eventually everything would be in precisely the same place. Is there any other possible conclusion?

As early as 1963, Russian scientists Evgenii Lifshitz and Isaac Khalatnikov proposed another possible ending to this time-reversed story. They ran the history-of-the-universe film backwards, imagining a scenario in which the universe contracts and all the galaxies draw closer to one another, appearing to be on collision course. But Lifshitz and Khalatnikov pointed out that the galaxies have other motion in addition to the motion that brings them towards one another. Could it be that this additional motion might cause them, as they approach one another, to miss one another and fly past with a nod, so to speak? Was this the way to avoid having a beginning? If we kept watching the movie backwards, might we simply see the universe expand again? Or must everything have begun in the same spot?

It was this question that engaged Stephen Hawking of Cambridge and Roger Penrose of Oxford in the middle and late 1960s.

Hawking's story is well known. He has become a legend in his time – the physicist who explores the frontiers of scientific knowledge and speculation, who works by spectacular leaps of intuition, who often seems to reverse himself, who writes best-selling books that readers struggle to understand. Motor neurone disease has locked Hawking's body motionless in a wheelchair, but he is the most nimble-minded of men.

Roger Penrose also thinks outside the envelope. In his youth he discovered an 'impossible object' – which means a figure that can't really exist because it contradicts itself. His father helped him turn the idea into the 'Penrose Staircase', and Maurits Escher used it in two of his famous lithographs: *Ascending and Descending* and *Waterfall*. Penrose also managed to visualize an impossible object in four-dimensional space. As he matured he never gave up 'playful' mathematics. He's become one of the world's most imaginative mathematicians, physicists and authors, and has discovered two shapes ('Penrose tiles') that in their three-dimensional forms may underlie a new kind of matter.

In 1965 Penrose, building on earlier work of John Archibald Wheeler, Subrahmanyan Chandrasekhar and others, showed that if the universe obeys general relativity and several other constraints, when a very massive star has no nuclear fuel left to burn and collapses under the force of its own gravity, it will be crushed to a point of infinite density and infinite spacetime curvature – a 'singularity'. General relativity predicts the existence of singularities. In the early sixties some physicists speculated that a star of great enough mass undergoing gravitational collapse might form a singularity at the centre of a black hole, but very few took this prediction seriously. Penrose calculated that this will happen even if the collapse isn't perfectly smooth and symmetrical. No 'might' about it. It must.

Hawking, in his doctoral thesis at Cambridge in 1965, reversed the direction of time and applied the same concept to the entire universe, suspecting that if he could watch the expansion of the universe run backwards he would discover something similar to what Penrose had found with black holes. Once the collapse (the expansion of the universe run in reverse) had proceeded far enough, whatever additional motions the galaxies had would make no difference to the history of the universe. By 1970 Hawking and Penrose were able to demonstrate, in Hawking's words, 'that if general relativity is correct, any reasonable model of the universe must start with a singularity'. Everything that was to be the matter/

energy of the universe that human beings might eventually be able to observe would have been compressed not to the sphere Lemaître envisioned (the primeval atom) but to something much smaller than that – to a point of infinite density.

Well, that did it! Physical theories can't work with infinite numbers. When the theory of general relativity predicts a singularity of infinite density and infinite spacetime curvature, it also predicts its own breakdown. All the theories of classical physics are useless at a singularity. There is no possibility of predicting what will emerge; one can only wait to observe what it will be. Indeed, why should anything emerge at all? There is no way to find out why this singularity suddenly ceases to be a singularity and becomes a universe. And what happened before the singularity? It's not even clear whether that question has any meaning.

Hawking and Penrose's discovery did not, however, put an end to attempts to devise an origin-of-the-universe story that would be more palatable to those unwilling to accept impenetrable obstacles and hints of a Creator. Hawking would soon be one of those most eagerly trying to untie the Gordian Knot he and Penrose had discovered.

CHAPTER 7

Deciphering Ancient Light
1946–1999

When people on airplanes ask me what I do, I used to say I was a physicist, which ended the discussion. I once said I was a cosmologist, but they started asking me about makeup, and the title 'astronomer' gets confused with astrologer. Now I say I make maps.

Margaret Geller

In the second half of the 20th century, the cosmic distance ladder became increasingly sturdy. Measurements using old and new techniques served as checks on one another, previous estimates were re-examined with improved technology and fresh theoretical understanding. It became possible to discern as never before the structure of the Galaxy and of the universe as a whole.

In 1946, researchers had first bounced a radar signal off the Moon. After that, astronomers were able to fine-tune distance measurements within the solar system by bouncing signals off the planets and the outer atmosphere of the Sun and timing how long it took the radar echo to come back. Later, unmanned missions visited the planets. If you send a spacecraft to Mars, and it gets there with only minor course corrections along the way, you know you have Mars's distance and orbit more or less right. That settles any argument between you and Cassini.

Between 1838 – when Bessel, Henderson and von Struve first measured stellar parallax – and 1900, astronomers used the parallax method to measure approximate distances to no more than about 100 stars. Because the Earth's atmosphere refracts light rays passing through, blurring the images of stars, the best ground-based telescopes, even today, measure accurate parallaxes of the brightest stars out to only about 300 light years, a minuscule distance by the standards of the Galaxy. The potential for considerable extension of that range came in the late 1950s with the ability to put telescopes beyond the atmosphere. By the early 1990s, astronomers had reasonably accurate parallaxes for close to 10,000 stars. In 1989, the European Space Agency launched the satellite Hipparcos – or High Precision Parallax Collecting Satellite. Hipparcos still measures parallaxes from the base line provided by the Earth's orbit, but it can measure them several times further away than earth-bound telescopes, in a volume of space 100 times as large. By the mid-1990s, Hipparcos had swelled the catalogue of precisely measured distances to 120,000 stars. Before it came on line there was still no more accurate way to determine the distance to the nearest star cluster, the Hyades, than the old 'moving cluster method'. Hipparcos measures its parallax directly.

Historically, one of the most important advances in cosmic measurement was finding that Cepheid variables could serve as 'standard candles', distance calibrators to measure beyond the parallax range. However, when it came to finding absolute distances, rather than how the distance of one star relates to another, this rung in the cosmic distance ladder depended on there being Cepheids close enough to be measured directly by parallax. There were none. The best that could be done was to measure their distance by a version of statistical parallax, which meant that the Cepheid rung was a less reliable part of the ladder than it might have been if there were closer Cepheids. Again the Hipparcos satellite finally promises a breakthrough. It has now measured a few Cepheid parallaxes directly. There is some question about the reliability of these measurements, but astronomers hope these uncertainties will be resolved by a project called the Space Interferometry Mission in the first decade of the 21st century.

Experts have studied the physics of stars, the rate at which they convert hydrogen to helium and burn up their fuel, their masses, their temperatures, their life cycles, their spectra, the way they affect one another in binary and ternary groupings, the idiosyncrasies that set some of them apart, and the wobbling that indicates there are planets orbiting a star. These days, the child of astronomer parents who sings the nursery song 'Twinkle, twinkle, little star. How I wonder what you are!' is asking for an extraordinarily long lecture! But none of this burgeoning knowledge about stars has diminished the significance of Cepheids. They are the best hope for establishing absolute distance measurements far into the universe. The Cepheids Leavitt first studied were in the Magellanic Clouds, about 169,000 light years away. With modern ground-based telescopes astronomers detect Cepheids in galaxies 15 million light years distant. With the Hubble Space Telescope, NASA's orbiting observatory that came into full use after repairs in December 1993, researchers study them in galaxies near to 60 million light years away. The Hubble telescope's mirror is smaller than many ground-based telescopes, but because it's based outside the distorting atmosphere of the Earth, it concentrates the light from stars into images that are many times sharper.

Cepheids were only one of the identifiable families of stars that astronomers found among the growing sample whose distances they could measure or estimate using various parallax techniques. One other valuable category are the RR Lyrae stars, which all have the same absolute magnitude averaged over each star's varying brightness.

Looking at stars' spectra has proved to be among the most effective ways to gain knowledge about them. In the past few decades astronomers have continued to study the spectra of thousands of them whose distances are known from various methods. Working from a discovery of Hertzsprung's that for stars of any particular spectral type there is a correlation between the width of the spectral lines and a star's absolute magnitude, they have succeeded in compiling tables giving the absolute magnitude for stars with *any* combination of spectral type and line width. Check a star's spectral lines, look up the absolute magnitude it should have, compare this with its apparent magnitude, and you know its distance! The method is called spectroscopic parallax. Its precise accuracy depends on how correct the measured distances were for the stars used to compile the table, but astronomers believe it is, for the most part, highly reliable. Not only does this method allow one to measure the distance to any star for which a spectrum can be obtained, it also makes it possible to find the distance to a cloud of gas or dust by finding a star in it and measuring that star's distance.

Another tool that, like spectroscopy, was inherited from the 19th century and put to optimum use in the 20th is the Doppler shift. When Slipher measured the red shifts of the nebulae in the 1920s, he used the 24-inch telescope at the Lowell Observatory, which was one of the finest of that day. But Slipher had to spend night after night in the unheated telescope dome, exposing each photograph for 20 to 40 hours to obtain spectra from which he then could measure the shifts. The equipment used by Edwin Hubble was better, but he and his contemporaries still had to expose a single photographic plate to hours of light coming through a telescope to record the tiny spectrum that they could then study with a magnifier to find the spectral lines. Today the job of finding the red shifts of galaxies and quasars is done in minutes by telescopes equipped with charge-coupled devices (CCDs) – silicon chips that convert light from the night sky into digitized images. Computer-run arrays make it possible to take the spectra of many galaxies at once.

Red shift has become the tool of choice, and often also the tool of necessity, for measuring the distant universe, and most of the mapping on the largest scales has been based on red-shift measurements only. By mid-century, astronomers had measured red shifts for about 100 galaxies. By 1970 the number was around 2,000. Today, red shifts have been catalogued for more than 100,000 galaxies and counting. There are ongoing projects such as the Sloan Digital Sky Survey which alone is expected to collect the red shifts of one million galaxies.

Red-shift measurement is not, however, infallible. How much the light from an object is red-shifted gives an indication of how rapidly the object is increasing its distance from Earth, providing a basis for comparing the distances of far-off objects. The method would work flawlessly in an ideal situation where everything was moving directly away from everything else with the expansion of the universe (like the raisins in the rising loaf of bread) and there was no other motion going on. Unfortunately, the situation is not that simple.

Galaxies do more than recede from one another with the expansion of the universe. Some of them are locked in binary systems, with two galaxies orbiting a common centre of mass, so that at any one time, one of the pair is likely to be moving away from us at a speed greater than the simple expansion of the universe and the other at a speed less than that expansion rate, though they are really about the same distance away. Spiral galaxies also spin like giant Catherine wheels, and if a binary system is made up of two spiral galaxies, they probably spin in opposite directions, causing yet more complication in the overall motion. Pairs and singles among galaxies are almost all grouped together in galaxy clusters, orbiting the gravitational centre of the cluster. The clusters orbit the gravitational centres of superclusters, and all of these – galaxies, clusters and superclusters – are pulling and tugging at themselves and one another by means of their gravitational attraction. Since it's extremely difficult to take all of this motion into account and interpret it correctly, measurements derived from red shift are continually subject to fine-tuning.

Halton C. Arp, who has spent many years at Mount Wilson Observatory studying galaxies, has cast a minority vote on the question of how much red shift should be trusted as a way of measuring distance to far-off galaxies and quasars. With Geoffrey Burbidge of the University of California, Arp has studied the brightest quasar, 3C273. Its red shift indicates that it should be about two billion light years away. Arp has found, however, that 3C273 appears to be interacting with a giant elliptical cloud of hydrogen gas no more than 65 million light years away in the constellation Virgo. There are other mysterious cases where objects at dramatically different red shifts appear to be linked. Arp believes that quasars are not as remote as most astronomers have been measuring them, but much closer to home. Most of Arp's colleagues pass off his evidence as coincidence.

While astronomy was probing deeper and deeper into space as the 20th century progressed, at a rate that makes earlier centuries look somnolent by comparison, it was also meeting some frustrating obstacles. Within the Milky Way Galaxy, visibility with the naked eye and optical telescopes is obscured by blotchy clouds of interstellar dust, especially in the direction of the Galactic centre. Of course, as Jansky discovered, the clouds of dust that block visible light are transparent to other types of radiation, but, except for radio waves and visible light, no other radiation can pass through the Earth's atmosphere. The revelations of early radio astronomy led many people to suspect that there might be much more to the universe than any earthbound telescope would ever see. World War II radar had helped usher in the age of radio astronomy. It took another impetus from the arena of international politics to boost telescopes above the Earth's atmosphere.

Space-based astronomy began as a distant by-product of the Treaty of Versailles, which ended World War I. That treaty limited Germany to the production of artillery of small calibre. Germany turned research efforts and funds to rocketry instead of artillery and made significant progress in this new area. One result was the V-2 rockets used to attack southern England in 1944 and 1945, during World War II. A captured stock of V-2 rockets ended up in the United States, and 25 of them were set aside for scientific use. Projects were soon underway to develop controls that could position these rockets with sufficient accuracy and stability for astronomical observations to be made from them.

By the late 1950s, Western technology had improved steadily, and scientists were becoming accustomed to large budgets justified by the Cold War arms race and national security interests, even for projects that had no immediately obvious practical applications. A partnership of unprecedented scope between government and big-ticket science had begun. However, it was a surprise from the Soviet Union that jump-started Western space-age astronomy in earnest. On 4 October 1957, the Soviet Union launched Sputnik 1, the first human-made satellite to orbit the Earth. The perception was that the Soviet Union had pulled ahead in the race into space and, by implication, in the arms race as well. To catch up, the Western nations, particularly the United States, began to pour money into the development of technology for spacecraft and satellites, and also into science education. I was a teenager then, and I must admit I can't recall there *being* a race into space until suddenly 'they' were winning. The panic filtered down to the level of my high school in Texas which wasn't, evidently, as good at producing scientists as Soviet schools were. Better physics and maths books and lab equipment were purchased, teachers went to workshops for retraining, and we were all urged to take more classes in these subjects. My younger brother's stock, as a fledgling physicist and computer whiz, went up. Mine, as a classical musician, went down. At the University of Texas, my later-to-be husband was called unpatriotic by his maths professor for choosing not to major in maths.

Before long the Western nations also had sent spacecraft into orbit, and the race continued, to the enormous benefit of all astronomy, not just space-based projects. Cold War competition paid the escalating bills. Much had changed since 1609, when Galileo put together his own perspicillum, and the 19th century, when Lord Rosse financed the construction of his Leviathan and a university community pooled their resources to buy Harvard's first good telescope. In the early 20th century, a portion of Lowell's fortune was enough to build a major observatory, and a corporation, AT&T, paid for the telescope at Bell Labs. The cost of astronomy in the late 20th century was on another scale entirely, which strained even the largest national budgets.

Fortunately, those nations that had the wherewithal also had motivation, other than the furthering of human knowledge, for supporting science. The technology that allowed telescopes to fly above the atmosphere, pointing upwards, also made it possible for surveillance cameras to point downwards. The technology that built and guided space probes also built and guided missiles. Moreover, the world image of a modern superpower included, just as it had done on a smaller scale for the Medicis in Renaissance Florence and Louis XIV of France, an image of technological and scientific pre-eminence.

When telescopes and cameras probed for the first time above the atmosphere in the late 1950s and early 1960s, those who put them there found there was a great deal to be seen. The full potential of space-based astronomy took years to come to fruition and is still not fully realized. Nevertheless, a new era had begun. It was possible at last to view the universe without the refraction problem that had frustrated astronomers since Galileo and Tycho Brahe, and to study objects at many different wavelengths, comparing findings in one range with those in another. There were indeed things out there that don't radiate at all at optical or radio wavelengths and others that look very different when viewed in other parts of the spectrum.

Unfortunately for our understanding of the Milky Way Galaxy, even all of this improved vision did not provide an easy way to calculate the distances to hot nebulae and clouds of gas in which stars are born, and it didn't allow astronomers to see the Galaxy from the outside. Though they would in time find ways to compile maps that are almost as good as seeing the Galaxy from afar, it was still easier to study things beyond the Galaxy.

The most significant progress in that line of enquiry continued for a while to be made not from space but with ground-based telescopes. Hubble had used galaxies themselves as standard candles. Other astronomers had chosen to assume that the brightest galaxy in a major cluster has approximately the same absolute magnitude as the brightest galaxy in every other major cluster. This method turned out to have some pitfalls: if we see two candles at a distance from us at night, and know that the two would have identical brightness if we saw them close up, we can judge their distance relative to one another by how bright they appear to us – unless without realizing it we happen to catch one of the candles just as a moth has flown into it and caused it to flare unusually bright. There is a risk of something like that happening when researchers use the brightest galaxies in galaxy clusters as standard candles. In a crowded cluster, it isn't uncommon for galaxies' paths to cross or come near enough so that a large galaxy 'cannibalizes' a smaller one, making the cannibal for a time much brighter than normal. The 'temporary' change can last a few hundred million years. Observing a distant galaxy cluster, and deciding that what appears to be the brightest galaxy in it can be used as a distance calibrator, involves a risk of choosing a galaxy that has made a meal of another within the last few hundred million years and therefore cannot be fairly compared with the brightest galaxies in other clusters, who may still be hungry.

Addressing this and other problems, Allan Sandage with Gustav Tammann from Switzerland carried cosmic-distance-ladder construction to unprecedented heights. Beginning in the early 1960s, Sandage and Tammann began using the 200-inch Mount Palomar telescope to set in place carefully calculated rungs leading deeper and deeper into the universe. They first utilized Cepheids to estimate distances to galaxies in the 'Local Group', the group that includes the Milky Way, Andromeda and other relatively nearby galaxies. Their next step, also with Cepheids, got them to a great spiral galaxy, NGC 2403, in another 'group' called M81. With these distances in hand, they proceeded to measure the size and luminosity of huge clouds of ionized hydrogen gas in these galaxies, and to study how the size and luminosity of these clouds is related to the overall luminosity of the galaxies. That relationship gave them a way to calculate the distances of more remote galaxies containing similar gaseous clouds.

Sandage and Tammann also picked up on a new scheme from Canadian Sidney van der Bergh for classifying spiral galaxies. Van der Bergh estimated their luminosity from the clarity and contrast of the spiral arms. Here was yet another standard candle to gauge distances of even more remote galaxies. By 1975, Sandage and Tammann had concluded that the universe was possibly as old as 18 billion years – a considerable increase over the 13 billion years Sandage had estimated in 1958 and a highly satisfactory result, because it allowed the universe to be sufficiently old to encompass the age of its oldest stars, even the most ancient globular clusters.

Sandage and others also found supernovae increasingly useful as yardsticks. Supernovae are exploding stars and they are extremely bright, which makes them the easiest stars to observe at great distances. One estimate has it that in a one-minute interval, a supernova can put out more energy than all the 'normal' stars in the observable universe. Only a small part of this energy is in the visible part of the spectrum, but even that can be sufficient to outshine the entire galaxy in which the supernova occurs. It goes without saying that such events could be remarkable standard candles – if they are in any way standard. If, for instance, all supernovae reach the same maximum brightness, that would mean that the differences in maximum observed brightness are attributable only to distance, providing an excellent yardstick for finding out how the distance to one supernova compares with the distance to another. It might also become possible to understand individual explosions sufficiently well to estimate their distances even if each is unique and comparisons among them are unhelpful. There has been work on both these fronts, and the search for and study of supernovae far beyond the Galaxy is currently one of the ways researchers are most successfully pushing back the frontiers of astronomy and astrophysics.

It turns out that all supernovae do not reach the same brightness, but for a while astronomers thought one particular class, Type Ia, did. It was a setback in the 1990s to find there were brightness differences among Type Ia's. Fortunately those differences soon became well enough understood to restore Type Ia's to their status as standard candles. Finding the distance of other types of supernovae is still problematical, but analysis of emissions in different wavelengths enables experts to find what type of star has exploded and its mass, as well as to study detailed properties of the aftermath of the explosion. The blast typically produces radiation at many wavelengths as it expands and meets surrounding matter.

Supernovae fall into two broad groups called Type I and Type II. Type I supernovae are exploding 'white dwarf' stars. The story of such an event begins with an elderly star that has exhausted its nuclear fuel and collapsed to a sphere about the size of the Earth, with a mass probably close to or less than the mass of the Sun. When a star is that small and has that mass, the matter of which it is composed is packed to almost inconceivable density.

Many white dwarf stars don't lead solitary lives but are part of 'binary systems' in which two stars circle one other, orbiting their common centre of mass. Often a dwarf star's partner is a large, far less dense star. As they circle, the gravitational pull of the denser dwarf star cannibalizes matter from its companion and the dwarf gradually puts on weight (mass). Eventually it tips the scales at 1.4 times the mass of the Sun. That mass is called the Chandrasekhar limit, after Subrahmanyan Chandrasekhar. As a young theoretical physicist from India continuing his work in England, he calculated that limit in the early 1930s.

The dwarf star, having no more nuclear fuel to burn and having boosted its mass over the Chandrasekhar limit, collapses under the pull of its own gravity and rips apart in a titanic nuclear explosion. That cataclysm, observable far across the cosmos, is a Type I supernova. Since all white dwarf stars have approximately the same mass when they explode, astronomers reasoned that all these supernovae should have about the same absolute magnitude, which should make them good standard candles. They are that, in spite of some brightness differences between them.

Type I supernovae are relatively rare. In our Galaxy there was one in 1006, another in 1572 that Tycho Brahe saw, and another in 1604, observed by Johannes Kepler. Not a large sample. In order to find out whether Type I supernovae make good standard candles and the best ways to use them, it has been necessary to observe them in other galaxies whose distances are known. Fortunately, in terms of their energy output within the visible part of the spectrum, Type I supernovae are the brightest supernovae. Light from the most distant ones observed in the late 1990s has taken more than seven billion years to reach the Earth.

Type II supernovae, on the other hand, have never been dwarfs. They are exploding giants. The star that explodes in a Type II supernova is definitely well above the Chandrasekhar limit without having had to cannibalize a companion star. When an extremely massive star has used up all its nuclear fuel and can no longer support itself against the pull of its own gravity, it collapses and the resulting explosion is a Type II supernova. Though these are more powerful than Type Is, and can be detected at least a third of the way across the observable universe, they tend not to look as bright to the eye, because so little of their energy comes in the form of visible light. Since they range rather widely in their absolute magnitudes, Type IIs are unreliable as standard candles. However, the hope is that measuring the radiation from them, their temperature, and the velocity at which the stars' debris moves apart may provide a way to estimate their individual distances.

What is needed is the discovery of a great many more supernovae in order to put them to optimum use as distance indicators and perhaps find a way to measure their distances independently. Today, there are teams carrying on this search using the Hubble Space Telescope as well as earthbound telescopes in many parts of the world. Their work involves taking photographs of a large section of sky, away from the light of the Milky Way and nearby galaxies, then comparing these pictures with earlier ones of the same region. Computers scan the photographs, subtracting known galactic light, searching for any new light source. If a likely candidate shows up, researchers photograph the same area of sky later to see whether the light has moved. If it has, that probably indicates it came from a cosmic ray or asteroid. Examining genuinely new light sources in detail, astronomers look particularly for spectral patterns identifying those that are Type Ia supernovae – the type most useful as standard candles.

Timothy Ferris, in his book *The Whole Shebang*, tells of a far more grass-roots supernova search. The Reverend Robert Evans of the Uniting Church in Australia compares what he sees nightly through his telescope with his remarkable visual memory of the skies. Evans had discovered 27 supernovae by

1995, a record in the history of astronomy.

Studies of the still-expanding debris of old supernovae whose light reached Earth before our time indicate that the 1006 supernova was approximately 5,000 light years from Earth, and the 1572 supernova about 7,000 light years away. Tycho Brahe was right in insisting that this one was further away than the Moon. He called the phenomenon a 'nova', but as we now make the distinction a nova is a less drastic flare-up, usually of a white dwarf star, not an explosion that vaporizes an entire star once and for all. Astronomer Fritz Zwicky coined the name 'super-nova' in the 1930s.

Among the new techniques that have emerged in the last quarter of a century, there are several that measure the absolute magnitude of other galaxies, and some of these actually bypass the cosmic distance ladder. One is the Tully-Fisher Method, coming from American astronomers R. Brent Tully and U. Richard Fisher. They discovered in 1977 that the absolute magnitude of a spiral galaxy is related to something called its '21-centimetre line width'. Most of the interstellar matter spread throughout a spiral galaxy consists of hydrogen atoms, and these atoms emit radio noise at the wavelength of 21 centimetres. As a distant galaxy rotates (with some parts of the interstellar matter in it approaching us and other parts of it receding from us), the Doppler shift causes this spectral line to be blurred. How much it is blurred is directly related to the speed at which the galaxy is rotating. That speed is, in turn, related to the galaxy's brightness. It is possible to study the radio spectra at this wavelength from extremely faint sources.

Another new technique is the 'brightness fluctuation method', which consists of measuring the unevenness in the brightness of the surface of the central bulge of a spiral galaxy, or near the centre if the galaxy is elliptical. The reasoning is that there should be more apparent unevenness with nearby galaxies than with distant galaxies, because the nearer the galaxy the more easily it is resolved into stars and the less likely it is to appear as a smooth body of light.

Astrophysicists Zel'dovich and Rashid Sunyaev have discovered a third approach which measures the distance to faraway clusters of galaxies. They measure the intensity of the cosmic microwave background radiation *through* clusters of galaxies that are emitting X-ray radiation, and they find that the background radiation is heated up as it passes through such a cluster. The effect is a 'hot spot' in the background radiation. This radiation started out in the early universe with an extremely high temperature, and the more distant the intervening galaxy cluster, the more dense and overheated the 'hot spot' should be. By studying the hot spot, researchers estimate the distance to the galaxy cluster.

Understanding a fourth method, first suggested by the Norwegian Sjur Refsdal in 1964, requires some background information about 'gravitational lensing', a phenomenon that was predicted in theory early in the 20th century but not observed until much more recently:

According to Einstein, the presence of mass warps spacetime, and the greater the concentration of mass the greater the warp. Hence, when light travels through space, its path bends as it passes near massive bodies such as planets and stars, or concentrations of mass such as galaxies and galaxy clusters. There is a close-to-home example: the path of light from a star is bent as it passes near the Sun. If we were unaware of the effect, the position where the star appears (detectable during a solar eclipse) would cause us to estimate incorrectly the actual position of the star in the sky. See Figure 7.1.

It might seem that such distortion would more likely hinder than help measurement of distances to faraway galaxies. Not so. Take for example a situation in which light from a distant quasar is travelling across the universe towards Earth. Somewhere between the quasar and Earth is a cluster of galaxies, and the cluster is warping spacetime around it as such concentrations of

Figure 7.1

A beam of light travels from a distant star, passing near the Sun. The warping of spacetime near the Sun causes the path of the light to bend slightly inward towards the Sun. The Sun's brightness doesn't allow us to see such starlight in everyday circumstances, but during an eclipse, if we don't take into account the way the Sun bends the paths of light, it is possible to get a false impression about which direction such a beam of light is coming from and what the distant star's position is in the sky.

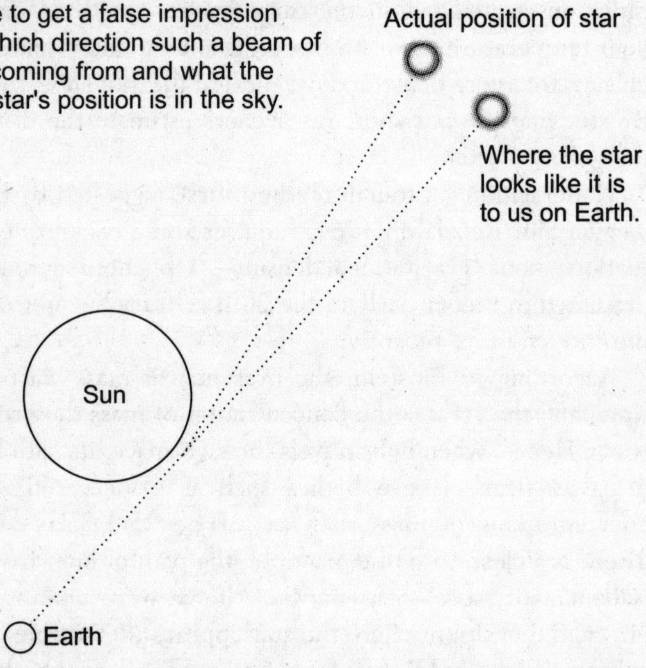

Actual position of star

Where the star looks like it is to us on Earth.

Sun

Earth

mass are wont to do. The warp acts as a lens, bending the path of the light from the quasar so that it reaches Earth via not one but two or even several paths. Observers on the Earth see two or more images of the quasar — or on occasion a ring of light — instead of a single point of light.

If the cluster responsible for the bending is precisely centred on a direct line between the quasar and the Earth, then the light passing around one side of the cluster will travel the same distance as the light passing around the other. But if the cluster isn't sitting dead centre, then the light coming around one side must travel further than the light coming around the other and it won't reach Earth as soon, since light nearly always travels at the same speed no matter what distance it must travel.

When light travelling from a quasar is lensed by a cluster of galaxies, studying the angles at which the light reaches Earth allows experts to calculate the extremely tiny fractional difference between the lengths of the paths. However, knowing, for example, that one path is longer than the other by one part in five billion doesn't give the actual length of the paths. Quasars have a characteristic that helps at this impasse. They flicker, flaring and fading irregularly over a span of days, weeks, or even years. When the quasar flares, observers on Earth see one image flare before the other, because that particular light has taken the shorter path. From the time delay between the flares, and how much the paths differ in length, it's possible to calculate the distance to the quasar. If, for instance, the time delay is one year and the paths differ in length by one part in five billion, the quasar is five billion light years away.

This method doesn't rely on any other steps in the cosmic distance ladder, even though quasars are some of the most distant objects from Earth. There are, in fact, none near us. Though they aren't as old as the cosmic microwave background radiation, they nevertheless beam their light to Earth like beacons near the limits of the observable universe. So much time has passed since the light we observe left the quasars that we see them as they existed when the universe was young, only a tiny percentage of its present age. By now, quasars have almost certainly evolved into something quite different from the way they appear to present-day observers. Perhaps into the sort of galaxies now situated closer to Earth.

In 1964, when Refsdal first suggested that gravitational lensing might be a clue to the distance of quasars, they were a very recent discovery and still something of an enigma. They looked more like faint stars than galaxies, but they also in some ways resembled gaseous nebulae. Those who studied these objects concluded that by cosmic standards they were very small but extremely bright, and their brightness varied over days, weeks, even years. All had astoundingly large red shifts.

Sandage was involved in some of the first investigation of these objects in the early 1960s, along with Thomas Matthews and Maarten Schmidt of the California Institute of Technology. At that time, existing physics was inadequate to make sense of them. Three possible ways of explaining their large red shift all presented problems:

First, the red shift might be caused by a gravitational field. We've said earlier that red shift is the result of the stretching of light waves as the object from which they are coming accelerates away from us. Gravity, like acceleration, stretches light waves, which isn't too surprising since even in more familiar situations, gravity and acceleration can feel the same and have the same effect. We experience it when the speeding up and slowing down of a lift makes us feel heavier or lighter, as though the pull of the Earth's gravity has changed. However, in the case of quasars, if it were a gravitational field causing the red shifts, quasars would have to be so massive and so near that they would disturb the orbits of the planets in the solar system. There was no sign that that was occurring, which seemed to rule out possibility number one.

Second, the objects might be stars in our own galaxy that have been ejected from somewhere with a force powerful enough to drive them away at a speed necessary for the measured red shift. Examination of the objects' spectra indicated that this was highly unlikely.

The best possibility left was that the objects are actually very distant indeed. To cause such a red shift as astronomers were measuring, quasars would have to be receding at as high a rate as 37 per cent of the speed of light. For such a red shift to be caused by the expansion of the universe they would have to be a vast distance away. How is it, then, that observers on Earth are able to see them as we do? For instance, one of the first studied was 3C273. Its red shift indicates that it is about two billion light years away. Yet it appeared as early as 1895 in photographs taken with optical telescopes of modest size. If it really is two billion light years away, it has to be radiating 100 times more power than the most luminous galaxies. That also seemed improbable to those who first struggled to understand these mysterious objects. How large *were* these things?

Their varying brightness was the giveaway. No source of light can flicker faster than radiation can cross it. If it did, the next flicker would begin before the previous one ended and the flicker would appear blurred, not a flicker at all. Radiation can't cross anything at a speed faster than the speed of light. The light output of 3C273 changes substantially in periods as short as one month, which means that most of the light from it must come from a region no larger than the distance light travels in a month.

Using this line of reasoning, researchers calculated that quasars are tiny by cosmic standards, but many have been discovered with greater red shifts than 3C273. For us to observe such small objects as we do, at such distances, they must be by far the brightest things in the universe, as bright as dozens or even hundreds of galaxies combined. Yet the light from 3C273 comes from a region only one light *month* across, while the light from a galaxy comes from a region of probably some 100,000 light *years* across or more.

Though they are much better understood now than they were in the early 1960s, quasars are still mysterious. The source of their power is probably a black hole at the core. Quasars are indeed all incredibly long ago and far away, some of the most remote objects in both space and time, and with their close relatives, violent BL-Lac objects and blazars, more powerful than any other sources of energy yet discovered.

As the century progressed there was a trend towards more comprehensive surveys and mapping of the Galaxy and the cosmos beyond. The first modern 'census' of the universe took place in the 1950s. It was the National Geographic Society–Palomar Observatory Survey, which relied on a 48-inch telescope and photographic plates.

By the mid-1950s, researchers were fairly sure that the Milky Way Galaxy was similar to thousands of other galaxies they were able to study at a distance, and that among them it was mid-sized. The closest Earthly observers come to 'seeing the Galaxy' with the naked eye is when we look up at the Milky Way in the night sky. That stream of light is part of the disc of the Galaxy, viewed edge-on (along the plane of the Galaxy), and not from outside the Galaxy but from within. Nevertheless, astronomers were able to glean from the study of other galaxies that a galaxy having a mass like the Milky Way has to be one of two types – a gas-free elliptical or a spiral like Andromeda. Ours is certainly not gas-free. That left spiral as the only option, which means – judging from knowledge of spiral galaxies – that it must have a Catherine wheel shape and be comprised of a thin disc of gas, dust and bright, relatively young stars; a central bulge of more densely packed older stars, around which the disc rotates; and a faint halo of even older stars. William Herschel compared the Galaxy to a grindstone. Modern astronomers wax even less poetic and compare it to a giant fried egg. The disc is the white of the egg, the central bulge is the yolk.

Beginning in 1950, American astronomer William Morgan of the Yerkes Observatory near Chicago pioneered the mapping of the Milky Way's spiral structure. He plotted the positions of two types of brilliant young stars – O and B stars, less than about 10 million years old – and found that they are arranged in two lines that run parallel to one another. One of the lines marks the arm in which our solar system lies, now called the Local Arm. The other line is the next arm out, the Perseus Arm. Morgan confirmed these findings by searching for nebulae and estimating their distances by studying the stars that light them. This investigation also revealed a third arm closer to the Galactic centre, later dubbed the Sagittarius Arm.

Other optical astronomers, following in the footsteps of Morgan, continued to study O and B stars and nebulae but ran into the problem of blotchy curtains of dust through which radiation in the optical range can't pass. This barrier is more like a forest in deep fog than an opaque wall, for in some areas it's possible for optical telescopes to penetrate through lighter dust, between what in a forest would be bushes and in the Galaxy are black molecular clouds. Nevertheless, in the optical range, astronomers hoping to explore nearer the Galactic centre can effectively see little further than 10,000 light years. Though the view is 50 times better looking in the other direction, from the southern hemisphere, curiosity about the centre of the Galaxy ran into frustration only about halfway there.

Meanwhile, Dutch astronomer Jan Oort had joined with Australian astronomers in a project to map the spiral arms using radio telescopes in Holland and Australia. Much of the interstellar matter spread throughout a spiral galaxy consists of hydrogen atoms, and these atoms emit radio noise at the wavelength of 21 centimetres. The study of 21-centimetre emission from hydrogen in the Milky Way Galaxy is one way of finding out how gas is distributed. Oort reasoned that the motion of the gas would show up as a shift in the wavelength it was emitting and would give him information about the way the Galaxy rotates. Plotting the speed of that rotation against the distance from the Galactic centre would enable him to measure the distance to gas clouds and nebulae.

Oort and his colleagues proceeded to map the spiral arms by

graphing the intensity of the radiation against the speeds measured from the red or blue shifts, reasoning that each peak in the graph was a hydrogen cloud and the shift revealed its distance. The resulting map of concentrations of hydrogen gas indicated a pattern of spiral arms, but the map didn't much resemble other spiral galaxies that can be seen more directly. It wasn't until the 1970s that the reason emerged. Hydrogen, it turns out, is not very strongly concentrated in the Galaxy's spiral arms, and some of the peaks on Oort's graph represented gas between the arms instead. Furthermore, the way gas moves and changes its speed upon entering a spiral arm leads to other misleading data. Hydrogen is not so effective a spiral arm 'tracer' as Oort hoped, though it did reveal the Outer Arm that lies beyond the Perseus Arm.

In 1976, Yvon and Yvonne Georgelin of France, who also had been following up on William Morgan's work, published a map of the Galaxy based instead on distances to hot nebulae. More recently, their map has been greatly refined by Patrick Thaddeus of Harvard and his colleagues, through the study of molecular clouds, the bushes in the optical forest. As with hydrogen, the motion of such a cloud shows up as a shift in the wavelength of radiation in the radio range of the spectrum – this time from carbon monoxide molecules. Except within a dense molecular ring near the Galactic centre – so crowded as to defy mapping of its overall structure – Thaddeus and his team have managed to separate individual molecular clouds along any line of sight by their different velocities, and from those velocities to calculate their distances from the Sun. The result is the most detailed and accurate map yet produced of the spiral arms.

Oort was on the right track to think that the best way to explore the structure of the Galaxy was to study the way things move in it. Most distant clouds of hydrogen or molecules of carbon monoxide have motion towards us or away from us, because the Galaxy doesn't rotate in the rigid way a solid structure like a wagon wheel would. What is true in the solar system is also true in a spiral galaxy: the influence of the gravity from the concentration of mass at the centre is weaker the further you are from the centre, and that means the outer parts of a spiral galaxy move at slower speeds, just as the outer planets in the solar system do. The result is that distant clouds don't stay continually the same distance from us. They do have motion – as much as 100 kilometres per second – towards us or away from us, and that motion shows up as a Doppler shift.

In both directions along the Sun's orbit around the Galaxy – the way we're headed and the way from which we're coming – the nearer gas clouds seem to remain always the same distance from us. Only further away in those two directions is there an observable shift. Looking from Earth in a third direction, directly away from the Galactic centre, there is virtually no shift. Looking in the direction directly towards the centre there is also no shift nearby, but in the vicinity of the centre itself there is violent non-circular motion of gas. On the near side of the centre the motion is towards us. On the other side it's away from us. Astronomers currently disagree about whether this movement indicates colossal explosive motion at the core of the Galaxy or means that the central bulge is bar-shaped.

Figure 7.2

Comparing the centre of a barred spiral galaxy with the centre of a galaxy having a circular central bulge.

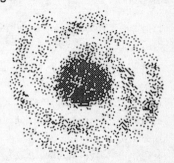

Most descriptions of the Galaxy as viewed face on, the direction from which it looks like a Catherine wheel, have the central bulge as circular. It may be necessary to alter that image slightly. Not all spiral galaxies have circular central bulges. Instead, about half appear to have a short bar at the centre with the spiral arms beginning at each end of the bar. (See comparison in Figure 7.2.) Observations by a satellite known as the Infrared Astronomical Satellite indicate that stars on one side of the Galactic centre are somewhat closer to Earth than stars on the other side. That would be explained if what is out there is a bar of stars set askew of our line of sight. (See Figure 7.3.) Also, the gravitational pull of the rotating bar could be causing the motions of clouds of gas observed near the Galactic centre. It may well be that the central bulge is only a short, fat bar, so that viewed from space beyond the Galaxy it looks more elliptical than round or rectangular.

There is an interesting glitch when researchers measure distances to an object closer to the centre of the Galaxy than the radius at which the Sun orbits, a glitch that illustrates the sort of complications modern astronomers encounter. Within this radius, discovering a particular speed of something from its Doppler shift doesn't give its distance. Instead, each speed indicates a choice between *two* possible distances. Along any line

Figure 7.3

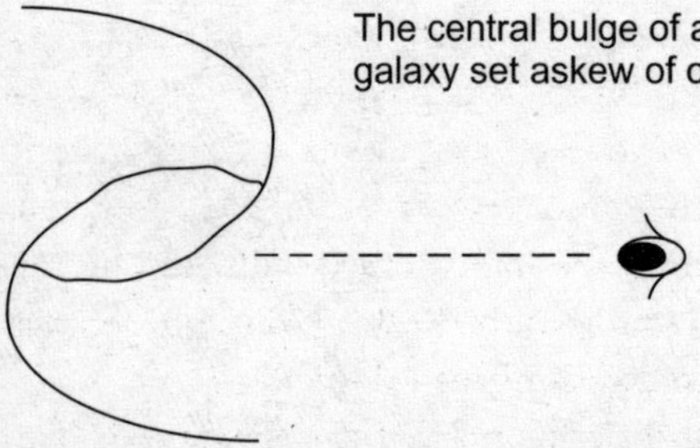

The central bulge of a barred spiral galaxy set askew of our line of sight.

of sight, the speeds increase with distance from us until we reach a point that is the same distance away as the Galaxy centre. Beyond that the speeds decrease again until they read zero once more for an object that is following the same orbit as the Sun but is on the other side of the Galaxy. It's necessary to look for additional clues to reveal whether something within the radius of the Sun's orbit lies in the near distance or the far distance – in other words, which of the two possible distances indicated by this speed is the correct one.

Illustration 16 (in the plate section) is one of the best images of the Galaxy as a whole, showing the disc and central bulge. It came from the Cosmic Background Explorer (COBE) satellite. This is not a direct snapshot but a map compiled from observations made at infrared wavelengths, a part of the spectrum that is not obscured by dust clouds or more diffuse interstellar dust.

While some astronomers have been exploring and mapping the Galaxy, others have been studying the universe outside it. One question such map-makers puzzle about is, could Friedmann have been wrong? Modern cosmology still accepts his assumption that the universe looks the same to us in all directions, and that anywhere we were in the universe it would look the same to us in all directions. And yet that isn't really the way it does look. Certainly not from Earth. In the sky from the southern hemisphere we see the Magellanic Clouds. We don't see them from the northern hemisphere, though from there we see another distant smudge of light, the Andromeda galaxy. When we look at the Milky Way we're looking along the plane of our Galaxy. We certainly see more stars there than in other parts of the sky. On this scale, anywhere we went in the universe we would likewise see visible matter distributed unevenly. Even on the much larger scale of a few hundred million light years, there is still structure including unimaginably large superclusters and voids 300 million light years wide. This is *not* evenly mixed raisin bread dough.

So if we agree with Friedmann that the universe is isotropic (looking the same in all directions) and homogeneous (with matter distributed evenly throughout space), we can't be talking about it on these scales. To find that it is isotropic and homogeneous, we have to look at the universe on a much larger scale yet. How large? In truth, astronomers have not yet found the level at which the universe would all look alike, where we would not be able to tell one sample of it from another. That has been one of the great surprises of the late 20th century.

The modern picture of the universe on the largest scales emerged in the 1980s and represented a dramatic change from the way it had been visualized before. Margaret Geller, John Huchra and Valerie de Lapparent, at the Harvard-Smithsonian Center for Astrophysics, decided it was worth following up on preliminary surveys that indicated there might be more structure in the universe than previously suspected. They proceeded to investigate by mapping the red shifts of a thousand galaxies across one strip of the northern sky. A strip becomes a wedge as we follow it further and further into space (see Figure 7.4), and this particular pioneering wedge now goes by the name of the Geller-Huchra Wedge.

When they began the project, no one including Huchra and Geller thought it likely to reveal anything particularly mind-boggling, and they were not even in any great hurry to interpret the data. When they did get around to it, jaws dropped. There was an eerie difference in this wedge of the universe from what nearly everyone had been expecting. Here was no homogeneous small-design-wallpaper pattern of galaxies with nothing to distinguish one portion of the picture from another. Instead there were huge voids with almost no galaxies in them, bordered by clusters of galaxies strung out like the lights of seaside towns and cities seen at night from a plane. There was a 'Great Wall' of galaxies, a billion light years long and tens of millions of light years thick. This was *structure*, to put it mildly! Huchra thought he and his colleagues must have made a mistake. Geller was more willing to relinquish old assumptions. As she quipped

Figure 7.4 The Geller-Huchra Wedge

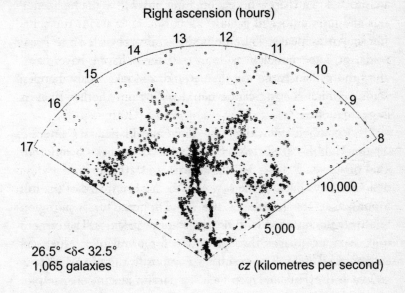

Right ascension (hours)

26.5° <δ< 32.5°
1,065 galaxies

cz (kilometres per second)

in an interview with science writer Timothy Ferris, 'I have a strongly held scepticism about any strongly held beliefs, especially my own.'

Geller and Huchra proceeded to investigate further by mapping the red shifts in the wedges of space to either side of the original Geller-Huchra Wedge, using more sophisticated equipment. There had been no error. The awesome structure went on and on.

Other astronomers (including Alex Scalay, David Koo, Richard Kron, T.J. Broadhurst, Richard Ellis and Jeff Munn) have since probed the depths of space by means of 'pencil beam' surveys. They limit themselves not to a strip of sky but to an area about half the size of the full Moon. We've said that a strip becomes a wedge as we move deeper into space. Similarly a small circle becomes a cone (as in Figure 1.4a). A pencil beam survey produces a cone-shaped three-dimensional map that keeps getting larger as it reaches further distances. The results are plotted in a chart where increasing red shift is the horizontal axis and the number of galaxies discovered at various red shifts are shown as peaks. This method has also revealed the great voids, or 'superbubbles' as some call them. None have diameters more than twice the diameter of the previously mapped voids – which is significant when questioning whether there is larger structure yet.

In December 1995, the Wide Field and Planetary Camera 2 on the Hubble Space Telescope was used to make a different kind of survey. It took 342 exposures over 10 consecutive days of a very tiny speck of the sky, $\frac{1}{140}$ the apparent size of the full Moon, near the handle of the Big Dipper. This is a region relatively uncluttered with nearby stars or galaxies. The camera took separate images through filters for ultraviolet, blue, red and infrared light. The resulting composite picture is what is known as the 'Hubble Deep Field', a narrow, deep 'core sample' of the sky, something like the core samples geologists take of the Earth's crust. The Deep Field doesn't reveal the ages of the galaxies it photographed or their distances. Galaxies at many different stages of the universe's history are stacked against one another in the picture.

The 'Field' reached back some 10 billion years and captured an unprecedented view of young, never-before-observed galaxies, some of which are four billion times fainter than can be detected by the human eye. Only the Cosmic Background Explorer satellite, when it measured the wrinkles in the cosmic microwave background radiation, looked deeper into space and into the past than this. The Deep Field showed spirals, ellipticals and a rich assortment of other galaxy shapes and sizes in many stages of evolution. Even though the photograph gives no reliable measure of the distances of the galaxies in it, the great number of very faint galaxies led astronomers immediately to suspect that some of them may have formed when the universe was very young. Researchers were soon hard at work measuring red shifts. By the spring of 1997, their calculations had yielded distances for thousands of the faint galaxies as far out as those having a red shift of 1. A red shift of 1 indicates that the light now reaching Earth from that galaxy left the galaxy when the universe was no more than half its present age. Larger red-shift numbers mean the light originated even further back in time. A red shift of 3, for example, means the universe was between 12.5 and 25 per cent of its present age.

Studying the galaxies as far out as those having a red shift of 1, researchers interpreted what they found to mean that galaxies with spiral and elliptical shapes probably have relatively uneventful lives (as galaxy biographies go). It seems that spirals and ellipticals must not change much over billions of years, for the oldest don't look markedly different from those nearby, and the number of them in the distant past was comparable to the number in the universe today. Experts hoping the Deep Field would reveal the formation process of spirals and ellipticals were disappointed. Evidently, to see that, they would have to push the search still further into the past.

Other types of galaxies in the Deep Field appear to have led more action-packed lives. Their irregular convoluted shapes suggest that galaxy collisions and mergers were far more common in the early universe than they are today, which makes sense, if things were so much more crowded together then. Researchers have also concluded from the Deep Field that the star formation rate in the universe has declined dramatically during the second half of its history.

For most of the galaxies found at the far limits of the Hubble Deep Field observations, it isn't yet possible to use red-shift measurements to calculate distances, because the amount of light from these faint galaxies isn't sufficient for even the largest telescopes to measure their red shifts. Astronomers have, however, worked out other techniques. One method is to use other objects as distance calibrators. Fortunately, some galaxies generate powerful emissions in the radio part of the spectrum, detectable at extremely great distances. Present understanding of these 'radio galaxies' has it that this emission comes from their active cores. Also fortunately, many of them are surrounded by other types of galaxies, and measuring the red shift of the radio galaxies allows astronomers to estimate how far away these whole clusters of galaxies are. Some are at red shifts as large as 2.3, and that means that the light reaching Earth today left them when the universe was less than 30 per cent of its present age.

State-of-the-art astronomy coordinates data from different kinds of telescopes – ground-based and space-based – observing in different parts of the spectrum. Some distant clusters of galaxies have been studied using the Hubble Telescope in conjunction with the most powerful ground-based telescopes, such as the Keck 10-metre telescope in Hawaii, and orbiting X-ray telescopes such as the German ROSAT X-ray Observatory. From this investigation astronomers have learned that some of the 'young' galaxy clusters were probably already extremely massive. Light from them shows that their stars were already mature when that light left them, so the galaxies must have formed much earlier.

Another approach to finding the distances of the faint galaxies in the Hubble Deep Field has been to study the light coming from quasars that are even further away, looking for evidence in the spectral lines that this light has encountered clouds of gas in halos around galaxies in its journey to Earth. Such detective work has nosed out galaxies at red shifts of 3 and even higher. Another helpful clue has been that all very remote galaxies have a distinctive 'signature' in their colour. Hydrogen, which is present both in galaxies and in the space between them, absorbs all ultraviolet light shorter than a certain wavelength. The upshot is that in the spectrum of light from the most distant galaxies there is a cut-off at that wavelength. Using filters, researchers find that a galaxy 'disappears' at wavelengths beyond that.

As of 1998, the most up-to-date maps of the universe continued to show clusters of galaxies lining up to form filaments enclosing vast voids. Clusters and superclusters are parts of supercluster complexes or 'walls' or 'sheets' up to a billion light years in length, enclosing even more enormous voids. Someone has commented that if we step back from this picture, we will see that the universe is a Swiss cheese. Richard Gott and colleagues at Princeton prefer a different mental image. Not a Swiss cheese, though that's close. A sponge, they say, is better. All of the material in a sponge is joined together. So are all the holes. In the universe the rich clusters of galaxies are more likely to be found at junctures the equivalent of where the pieces of a sponge's material come together. Other theorists think that galaxies, clusters and superclusters are like a glowing froth on an ocean of dark, invisible matter, similar to the froth that appears among ocean waves. Anyone who has swum in the sea has seen the voids, filaments and clusters of foam that form and reform continually on the swells.

In the existing three-dimensional maps, such as Geller and Huchra's, the largest structures visible are about the size of the volume surveyed. To get a good statistical sample of these enormous structures and to find out if there are even larger structures, it will be necessary to map larger volumes.

Two projects are now doing just that. A British and Australian project called the 2DF Galaxy Red Shift Survey uses the 3.9-metre telescope at the Anglo Australian Observatory on Siding Spring Mountain in Australia. By 2000 they hope to have measured a quarter of a million galaxy red shifts. Already they are seeing indication of the walls and voids. James Gunn at Princeton has been heading another new project, the Sloan Digital Sky Survey. Even in today's science world, where teams are the norm, SDSS is a remarkable conglomerate. It includes astronomers at the University of Chicago, Princeton, Fermilab near Chicago, Johns Hopkins, the Institute for Advanced Study, the US Naval Observatory, the University of Washington and the Japan Participation Group.

The Sloan Telescope itself, at the Apache Point Observatory in the foothills of the Sacramento Mountains north of El Paso, Texas, is modest in size but, linked to an impressive array of technology, it is powerful enough to undertake the largest and most comprehensive census of the visible universe that has ever been attempted, a survey of a quarter of the northern sky. Part of the Sloan apparatus consists of CCDs or 'charge-coupled devices', silicon chips that convert light from the night sky into digitized images that can be poured on to magnetic tapes and into computers. The project began surveying the sky in the summer of 1998, after nine years of organization, design and construction, and is expected over the next six years to produce images in five colours of 50 million galaxies, 100,000 quasars, millions of individual stars in the Milky Way, and other assorted items that might turn up. 'A field guide to the heavens', Mike Turner (one of the Chicago contingent) calls it. More than a field guide. A three-dimensional scale model of a large part of the universe.

Even when these projects have succeeded in mapping the universe in three dimensions, the situation will still be not unlike that in the 17th century when astronomers had cata-logued and charted what was 'out there' with greater and greater precision but were waiting for a Newton to come along and discover the dynamics that would reveal *why* things should have turned out as they have.

Not that the ingredients that will go into that understanding are entirely unfamiliar. Gravity and relativity are part of the explanation. Quantum theory – the theory of the very small (atoms, molecules and elementary particles) – is also certain to be an essential ingredient, because as large as the clusters and superclusters are today, in the early universe the material of which they are made was compressed in an area as small as anything in the quantum world. Seeds of the most colossal structures were probably sown by quantum fluctuations at a time less than a tiny fraction of a second after the Big Bang. The patterns of the ripples in the early universe, found in the cosmic microwave background radiation, would have depended on the way interactions occurred inside the fireball before it even reached the size of an apple.

Chaos and complexity theorists – who study randomness, the borders between what is random and what is predictable, and the patterns that emerge out of what seems to be chaos – also believe their theories have something to say about the dynamics that formed the large-scale structure of the universe. It would be difficult not to notice the resemblance between the universe on the largest known scales, the 'sponge' level, and the graphs and fractals produced in chaos theory.

Perhaps the best way to put together an overall mental picture of the structure of the Galaxy and the large-scale structure of the universe, as state-of-the-art astronomy describes them, is to take an imaginary tour. At the start, travellers should remind themselves that light travels at a rate of approximately 186,000 miles or 300,000 kilometres per second. It takes light 1.3 seconds to travel from the Moon to the Earth, 8.3 minutes to travel from the Sun to the Earth, four hours to travel from the Sun to Pluto, and about 4.3 *years* to travel from the Sun to the nearest star. Four hours to 4.3 years is an enormous leap, but within the range that light from the Sun can reach within 17 years, there are only about 50 stars.

Whether the Galaxy is large or small is, nevertheless, a relative matter. Taking into account the entire universe would mean not only considering planets and solar systems, but things far smaller yet – microbes, molecules, atoms, quarks, neutrinos. Next to those, *you and I* are by no means small, and compared to us the Galaxy is huge beyond comprehension. But among galaxies, it is mid-sized. Compared with the largest scale of structure known in the universe, it is vanishingly tiny.

Ignoring the fact that no human has ever viewed the Milky

Way Galaxy from a distance, imagine that one portion of the tour does start outside the Galaxy, facing the great spiral, then approaches it like a moth attempting to fly through the blades of a fan. Our itinerary takes us through the disc, not the central bulge. First we pass through a 'layer' several thousand light years thick, probably consisting of extremely hot gas and faint elderly stars. This region has been difficult to study and, so far, has not been well explored. There is some uncertainty about what is there.

Next we come to a region called the main disc, about 2,000 light years in thickness. Here are many of the Galaxy's stars, and if we choose the right trajectory through the main disc, we'll pass through our own solar system. If our route doesn't take us through one of the 'arms' of the spiral but instead goes through what look like much more sparsely populated areas in between, we'll be surprised to find not as much empty space as we might have expected. It won't be like flying through a fan and missing the blades, because the 'arms' are nothing so solid and distinct as blades of a fan. The number of stars per cubic light year is not much greater in a spiral arm than in the areas between the arms. However, passing through an arm we would be more likely to encounter extremely brilliant massive, short-lived stars and glowing nebulae lit by the light of the young stars.

Continue travelling and we'll come to the thinnest layer of the Galaxy. If the Galaxy were a chocolate-mint wafer, the main disc would be the chocolate and the inner disc we're now entering would be the mint cream layer. This is a disc of gas and dust, only 500 light years thick, and it is the Galaxy's nursery. It houses the youngest stars and is the birthplace of new stars. The resemblance to the chocolate-mint wafer is not exact. This thin disc of gas doesn't end out where the main disc layers end (as the filling of a chocolate-mint wafer does) but instead stretches out beyond that to a distance more than a third again as far from the centre of the Galaxy. At its extremes the gas disc bends like the rim of a hat. At one side of the Galaxy the edge of the hat brim curves upwards. On the other side of the Galaxy it curves downwards and then, further out, back upwards.

Continuing our journey out the other side of the Galaxy disc we find the same layers in reverse. Compared to its diameter along its plane (looking at it edge-on), the Galaxy is very thin indeed. A chocolate-mint wafer is too thick to be an accurate comparison, not large enough in diameter in relation to its thickness. The Galaxy has more the proportions of a gramophone record.

One thing astronomers now know – to follow up on an earlier attempt to map the Galaxy – is that Harlow Shapley was right to conclude that globular clusters outline the extent of the Galactic halo, though there are also some of them well beyond it. The Omega Centauri cluster, a globular cluster that Ptolemy catalogued (though not as a cluster), that Halley recognized as a cluster, and about which Herschel spoke in superlatives – 'richest . . . largest . . . truly astonishing . . . whose stars are literally innumerable' – does indeed turn out to be extraordinary. It is the brightest, largest and most massive cluster in the Galaxy, with tens of millions of stars. There are some 100 billion stars in the whole Milky Way Galaxy, many of them congregated in the central bulge, the bright yolk of the fried egg. If our journey had taken us directly through the bulge, we might have seen, or even ended up in, a massive black hole that is suspected to lurk at the very heart of all that brightness.

The Milky Way, which is now behind us as we move further afield, is only one among many billions of galaxies, not all of which are the same size and shape. They range from about 10 million to 10 trillion times the mass of the Sun, but that includes the extremes. A typical galaxy (the Milky Way is one of them) is midsize. We first pass among the other members of the 'Local Group' of galaxies. This group measures about five million light years across and is rather flattened in shape.

The designation, 'group', is part of the terminology astronomers use to describe the structure hierarchy of the universe, though this 'hierarchy' is not rigid, and thinking of the universe this way ignores the rich diversity of its structure. There are, moving to larger and larger scales, first galaxies, then groups, clusters, clouds, superclusters and supercluster complexes or 'walls'.

'Groups' typically include three to six conspicuous galaxies and a number of smaller, dimmer ones. The Local Group is no exception. The Andromeda galaxy is its dominant spiral, with an estimated 400 billion stars. The Milky Way ranks second, and there is a smaller but still impressive spiral called M33. These three giants are all that an observer in, for instance, the Virgo Cluster would be able to see of our Local Group, but lesser galaxies in the group outnumber them ten to one, and there are probably others not yet observed because they're hidden from Earth by dust clouds within the Milky Way. Andromeda has two small companions, M32 and NGC205, both elliptical galaxies. The Milky Way holds court with a retinue made up of its two best-known companions, the Magellanic Clouds, which are irregular galaxies, and three other satellite galaxies. Two of these, the Carina galaxy and the Sextans galaxy, are dwarf galaxies that are spherical in shape. The Sextans dwarf, discovered in 1990, is a little further away than the Magellanic Clouds and the total luminosity of its stars combined is only about 100,000 times the luminosity of the Sun – less than some single stars in the Milky Way. A more recently discovered satellite galaxy, the Sagittarius galaxy, is the closest neighbouring galaxy found so far. First recognized in 1993, it's a dwarf that looks likely to be cannibalized by the Milky Way. It has already lost some of its outer stars to the Galaxy's gravitational pull.

The distance to the Andromeda galaxy is fairly well established at some two and a quarter million light years, but there is still dispute about the distance to M33, the third largest Local Group spiral. When Edwin Hubble first measured it in the 1920s, he estimated that M33 was about as far away as the Andromeda galaxy. Allan Sandage, however, has reinterpreted

Hubble's data on the Cepheids used for that measurement, employing more modern techniques, and has concluded that M33 is more like three million light years from us, well beyond the Andromeda galaxy. Astronomers who have studied the same Cepheids at infrared wavelengths disagree with Sandage. They estimate that the two galaxies, though further away than Hubble estimated, are, as he thought, both about the same distance from us.

Groups like our Local Group have no particular structure or shape; in fact, they are also known as 'irregular clusters' and their galaxies are a hodgepodge of all types. However, that doesn't mean their existence is a random occurrence, with all these galaxies just happening to be passing close to one another on their way to somewhere else. All the galaxies in the Local Group are bound together by mutual gravitational attraction, and all are orbiting a common centre of gravity. Andromeda and the Milky Way are, at the moment, approaching each other at a speed of 300 kilometres per second. Slipher's discovery of the blue shift in Andromeda's light was not an error. There could eventually be a head-on collision, but that is not something to write about on the walls of the underground yet. The event is still a few billion years in the future, and the merger of the two galaxies will take another several billion years after that to complete. In the end there will probably be, instead of two spiral galaxies, one huge elliptical galaxy. Then again, Andromeda and the Milky Way may only circle one another in a polite do-si-do and then move apart again. Which will it be? In 2005, NASA plans to launch the Space Interferometry Mission, a spacecraft carrying an array of telescopes capable of determining, among other things, the exact angle of Andromeda's approach.

Obviously, relationships among galaxies are not always placid. Andromeda seems to have stripped M32 of a good number of its stars, while M32 has in turn caused distortion in the spiral structure of Andromeda. NGC205 is twisted by the pull of Andromeda. Most astronomers also believe that the Small Magellanic Cloud is being torn apart, probably by the gravity of the Milky Way Galaxy. There is a long, narrow ribbon of hydrogen gas, streaming half of the way around the sky and seeming to begin in the large pool of hydrogen that surrounds the two Magellanic Clouds. This may be gas from the ripped Small Cloud, left behind along its orbit around the Milky Way.

We would have to travel several million light years outside the Local Group to get to the nearest galaxies beyond it. But at this point we can learn more about the large-scale structure of the universe not by travelling further away from Earth, but by moving to larger and larger scales. Bear in mind the 'hierarchy' of the large-scale structure, but also be aware that the universe is not nested as neatly as Russian dolls, with galaxies within groups within clouds, within clusters, within superclusters, within supercluster complexes. There is far too much complication out there for such a simplistic picture to be anything but a distortion.

The Local Group is not far outside the border of the Coma-Sculptor cloud, a large cloud which in turn lies near the outer limits of the Virgo supercluster. The Virgo cluster, which is huge compared with the Local Group, is the giant heart of the Virgo supercluster and about 60 million light years from Earth. It is an enormous swarm of thousands of galaxies and a lot of hot gas – a 'regular cluster' because it doesn't seem to have such a mix of galaxy types as are to be found in motley assortments like our Local Group. Instead, it contains more than 1,000 prominent galaxies centred on a pair of giant elliptical galaxies, with probably many more less prominent galaxies that astronomers haven't yet detected, but very few spirals.

There's little point in even trying to get a sense of the vastness described here. Probably we come closest when we are feeling most overwhelmed by our *lack* of ability to conceive of such size and how it compares with familiar distances. The box on p. 248 gives approximate relative size scales in the universe as calculated in the mid-1990s, but in truth such numbers are beyond our human capacity to comprehend. Only for the sake of rough comparison, therefore, here are some figures having to do with the larger scales: a typical 'group' of galaxies is a few million light years across; the Local Group measures about five million light years; a cluster may be 10 to 20 million light years in diameter, a cloud some 30 million light years, a supercluster 100 to 200 million light years.

The question remains: Has astronomy discovered the 'top' of the hierarchy, or does it all go on and on to larger and larger structure? Will the voids turn out to be parts of systems of supervoids? Are the superclusters grouped in clusters of superclusters? Will we ever find a level at which the universe is isotropic and homogeneous? Is there such a level?

There are tentative answers to these questions. Pencil beam experts probing six billion light years of space (one three-billion-light-year-long beam pointing out each side of the Galaxy) have found no structure larger than the superbubbles. At extremely great distances, clusters, superclusters and voids seem to be spread more or less uniformly, with about an equal number in any direction you look. Perhaps it is here, in the ultimate sponge, that there is isotropy and homogeneity.

Astronomy has not been able to reveal an unbroken chronology leading from the era of the cosmic background radiation through to the present. There is a gap in the history of the universe like the lost years in the life of someone with amnesia. In terms of the structure of the universe there is one window (that provided by the cosmic background radiation) into the early universe about 300,000 years after the Big Bang . . . and, the next anyone is able to know, there are the voids and supercluster sheets and walls, much closer to the present in time and space. This is the same universe, but no one would guess that from appearances alone. Finding out what happened in the invisible stretch in between is one of the challenges facing the next generation of physicists and astronomers.

The Relative Scale of the Universe – as calculated in 1995

These numbers don't indicate actual sizes, only rough relationships between size ranges, expressed as 'powers of ten'. To translate: if the exponent is a positive number, it tells you how many zeros are in the number. For example: 10^3 is 1,000 (three zeros); 10^4 is 10,000 (four zeros); 10^8 is 100,000,000 (eight zeros). That means that something designated 10^8 is in the range of 10 times as large as something designated 10^7 or 100 times as large as something designated 10^6. If the exponent is a negative number, it tells how many zeros are in the denominator of a fraction with 1 as the numerator: 10^{-3} is $1/1,000$ (three zeros); 10^{-4} is $1/10,000$ (four zeros); 10^{-8} is $1/100,000,000$ (eight zeros). That means that something designated 10^{-6} is in the range of 10 times smaller than something designated 10^{-5}, 100 times smaller than something designated 10^{-4}, and so forth.

human beings to elephants	1–10
largest cities	10^4–10^5
Moon	10^6
Earth	10^7–10^8
Sun	10^9–10^{10}
Earth's orbit	10^{11}–10^{12}
solar system	10^{13}
galaxies	10^{20}–10^{21}
Local Group	10^{24}
observable universe	10^{27}
human beings to elephants	1–10
limit of visibility of the naked eye	10^{-3}–10^{-2}
visible light waves	10^{-7}–10^{-6}
molecules and viruses	10^{-9}–10^{-7}
atoms	10^{-10}
atomic nuclei	10^{-14}

However, the one window to the time before that blackout does suggest homogeneity. Though the most interesting recent news about the cosmic microwave background has been the discovery of a minuscule *lack* of homogeneity, it *is* remarkably homogeneous, and it is a picture from further in the past than any other we have. However, it shows the universe only in its extreme youth. It's like the smooth-skinned glamour shot that shows what a wrinkled dowager looked like in her late teens but gives us barely a hint of the 'structure' in that face today. It isn't evidence that the universe now is isotropic and homogeneous.

No human being, and no universe, can be captured adequately in a still photograph. The question of what is out there can't be considered meaningfully without asking, as well, how what is out there is moving. What movement does the Galaxy have, for instance, in relation to the rest of the universe besides Andromeda? One way to find out is to measure its motion against the cosmic microwave background radiation, because this radiation comes from a distance far beyond the remotest galaxies. The reasoning is that as the Galaxy moves through the cosmic background radiation, the radiation will measure warmer in 'front' of the Galaxy (in the direction towards which it's moving) and cooler 'behind'. If the temperature of the radiation reads the same in all directions, then the Galaxy isn't moving. George Smoot and colleagues measured these temperature differences in the early 1970s from high-altitude U2 planes. They found that the Milky Way Galaxy is moving at a rate of about 600 kilometres per second in the direction of the Virgo cluster. However, the Virgo cluster meanwhile is moving in the other direction and the distance between us is increasing – though not as rapidly as it would were the expansion of the universe the only movement at work here.

The motion of other galaxies besides our own is more difficult to plot, but one of the more intriguing results of this effort has been the discovery that several hundred galaxies, the Milky Way among them, are sidling off in a direction and at a speed that the expansion of the universe does not explain. The Great Attractor was the name Alan Dressler, one of those who first calculated this movement, gave to the huge theoretical mass concentration that must be drawing them, but it was difficult to identify any obvious culprit whose gravitational pull was to blame. The identity of the Great Attractor remained for a time one of the baffling enigmas of science, and the mystery has not yet been completely solved. However, in early 1996 Renee Kraan-Korteweg, at the Observatory of Paris-Meudon, and her colleagues reported sighting a massive galaxy cluster that appears to be located just about where the Great Attractor ought to be. Because dust in the Milky Way blocks much of this cluster's light, scientists had not realized before how wide and massive it is.

Does the universe as a whole move? Of course the universe is expanding, but what about other types of motion? For instance, does the universe rotate? Is it moving through some larger environment? The reply comes in the form of another question: In relation to what could the universe as a whole be said to rotate or not to rotate, or to move or not to move? We have come almost full circle to questions that date back to antiquity. Some of them have become meaningless. Others seem likely to remain forever unanswered. But some of the answers that our ancestors, even those with minds like Kepler and Newton, probably didn't think could ever be known by human beings are now, according to modern astronomers, astrophysicists and theoretical physicists, *almost* within our reach.

CHAPTER 8

The Quest for Omega
1930–1999

When I started, cosmology was very much like philosophy. There was very little chance of measuring something precisely. It's now turning into high precision science.

Alexander Szalay

How old is the universe? What is its future? Much of the work going on in state-of-the-art physics and astrophysics these days focuses on these two fundamental questions. In order to answer them, we must know the mass density of the universe – the elusive 'omega'.

The mass density of the universe means the amount of matter there is per cubic metre, averaged throughout the observable universe. Obviously this matter is unevenly distributed, at least on scales normally accessible to us. One way to find out the average density would be to add up all the matter in the universe and then divide by the number of cubic metres in the universe. On the face of it, that would appear to be a ludicrously difficult undertaking.

Never underestimate modern astrophysicists. It is possible to arrive at some estimates. In one method, called 'representative sampling', researchers divide the sky into sections of equal size, count the number of galaxies in a section, then multiply the count from that section by the total number of sections. Combined with knowledge about the masses of galaxies, this procedure gives a rough estimate of the total mass of the universe. Studies like the Hubble Deep Field make such sampling increasingly substantive.

Another way to try to find the average mass density is to study the way the universe appears to be working – how fast it's expanding, whether the expansion is speeding up or slowing down, how gravity seems to be affecting different parts of the universe, what other forces besides gravity come into play, and how the contents of the universe have evolved over time. This sounds considerably more difficult than counting galaxies and multiplying by sections. It is extremely complicated. Nevertheless, theorists have a formula that they believe shows how the mass density of the universe is related to such questions, an equation that allows them to weigh the answers one against the other. It is the so-called 'equation for omega'.

The equation shows how the mass density, omega, affects the future of the universe. If omega turns out to be more than one (meaning that there is more than an average throughout space of one hydrogen atom per 10 cubic metres), the universe will eventually stop expanding and contract. That would be a 'closed' universe. If omega is less than one (less than an average throughout space of one hydrogen atom per 10 cubic metres), the universe will expand forever. That would be an 'open' universe. If omega is precisely one, then the universe is at the 'critical density' that will allow it to expand at precisely the right rate to avoid recollapse, eternally slowing down its expansion but never completely ceasing to expand. That would be a 'flat' universe – the type of universe inflation theory predicts. (See box below.)

Why should there be such a tight connection between the mass density of the universe and the fate of the universe? First,

Omega is *more* than one . . . universe eventually stops expanding and collapses. (A 'closed' universe.)

Omega is *less* than one . . . universe expands forever, thinning out eternally. (An 'open' universe.)

Omega *equals* one . . . critical density; universe expands at precisely the right rate to avoid recollapse. (A 'flat' universe, the sort of universe predicted by inflation theory.)

'mass' is the measure of how much matter there is – in a planet or star or galaxy or, in this case, in the universe as a whole. Every particle of matter in the universe is attracting every other by means of gravitational attraction. How greatly objects are influenced by one another's gravitational attraction depends on how far apart they are. The closer they are the more they 'feel' one another's pull. So when it comes to the question of whether or not the universe will eventually contract or whether it will keep expanding, much hangs on how densely or thinly the matter in the universe is spread out. In fact, the mass density very possibly does dictate the fate of the universe.

Solving the equation for omega requires knowing four numbers, three of which are currently not known with certainty. The four are the speed of light, the cosmological constant (the theoretical constant Einstein suggested and that he hoped would allow the universe *not* to expand or contract), the Hubble constant and the deceleration parameter. The third and fourth need introduction:

The *Hubble constant* or H_o (pronounced 'H nought') denotes the rate at which the universe is expanding. However, it isn't a direct indication of the speed at which everything out there is rushing away. To give a specific example: if the Hubble constant is 50, that indicates that there is an *increase* in the recession velocity of 50 kilometres per second for every megaparsec of distance from the observer doing the measuring. Taking the simplest case and remembering that there *are* no receding galaxies this close to Earth, if the Hubble constant is 50 and Galaxy A is one megaparsec away from Earth, Galaxy A should be receding at a velocity of 50 kilometres per second. If Galaxy B, out beyond Galaxy A, is two megaparsecs away from Earth, Galaxy B should be receding at a velocity of 100 kilometres per second. A galaxy three megaparsecs away . . . 150 kilometres per second, and so forth. This also means that if the Milky Way, Galaxy A and Galaxy B were all lined up in a straight line, and if you and I were in Galaxy A, we would find Galaxy B receding at a rate of 50 kilometres per second in one direction and the Milky Way Galaxy receding at 50 kilometres per second in the other. Notice that this does follow the raisin bread rule of twice as far, twice as fast, no matter what value the Hubble constant has, and no matter where you and I stand to watch other galaxies recede.

The *deceleration parameter* measures the rate at which the expansion is slowing down due to the gravitational attraction among all the clusters of galaxies. It might seem astrophysicists ought already to know what that is. If it's true that when we observe very distant galaxies and galaxy clusters we are seeing them as they were billions of years ago, why isn't it possible to compare the rate at which they are receding with the rate at which nearer galaxies are receding, find out whether the expansion has slowed down and, if so, how much? Astrophysicists are trying to do just that, but it isn't easy. Some have suggested that one reason why it's difficult may be that the universe *is* perfectly balanced between the mass density that would allow it to expand forever and the mass density that would cause it to collapse to a 'Big Crunch'. In other words, perhaps the very difficulty of making that determination is a clue that omega equals one and the universe is expanding at precisely the right rate to go on forever, always slowing down its expansion but never stopping the expansion and collapsing.

One source of complication is that the mass density of the universe is changing over time. Unless new matter or energy is appearing on the scene (which the Steady State theory proposed but most physicists don't believe is happening) things inevitably grow less and less dense in an expanding universe. They thin out.

It is the interrelatedness of these numbers, or values, that's laid out concisely in the equation for omega. It doesn't take much expertise to see that there are relationships, that one thing depends on another. The equation shows precisely in what manner and to what degree they are related. At the risk of sending a great many readers running for cover, here (Figure 8.1) is the formula for omega. Consider it a souvenir, something a patient reader is owed for having made it so far with this book! We will *not* proceed to solve it.

How far have researchers got in the process of discovering the unknown numbers in that equation? They know the speed of light. What a pleasure to be able to plug in one actual number here! No one yet knows the value for the deceleration parameter or the cosmological constant. There is disagreement over

Figure 8.1   The Formula for Omega

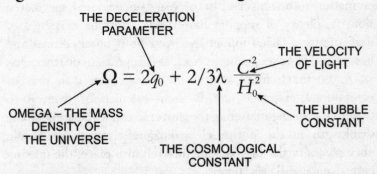

THE DECELERATION PARAMETER

THE VELOCITY OF LIGHT

$$\Omega = 2q_0 + 2/3\lambda \ \frac{C^2}{H_0^2}$$

OMEGA – THE MASS DENSITY OF THE UNIVERSE

THE HUBBLE CONSTANT

THE COSMOLOGICAL CONSTANT

the Hubble constant. Modern scholars find themselves in very much the same situation their forebears were in when they had Kepler's laws but not Cassini's and Flamsteed's measurements of the distance to Mars. Here is a formula – but not all the numbers to put into it.

One serious problem in estimating how much matter there is in the universe is that there actually doesn't seem to be enough of it around. In the 1930s, Swiss astronomer Fritz Zwicky discovered that galaxies in the 'great cluster' in the constellation Coma Berenices were moving too rapidly, relative to one another, to be bound together by their mutual gravitational attraction. Given the way gravity works, and what can be seen of this cluster of galaxies, the arrangement should be flying apart. Searching for an explanation, Zwicky thought of two possibilities. What appeared to be a cluster might instead be a short-term random grouping of galaxies; *or* there might be more to these galaxies than met the eye or the telescope. In order to provide the amount of gravitational attraction required to bind the cluster together, it would have to contain much more matter than we observe. No one was willing to consider the third possibility that physicists might have made an egregious error in figuring out how gravity operates, or in assuming that it operates the same everywhere.

With Zwicky's discovery was born the puzzling notion that it may be impossible to observe more than a tiny fraction of all the matter in the universe. In the years since he first speculated about it, plenty of support has emerged for the existence of 'dark matter'. That support has been both observational and theoretical. For everything to work as it appears to do, there has got to be much more matter in the universe than present technology is able to detect. By some calculations, from 90 to 99 per cent of the matter in the universe is not radiating at any wavelength in the entire electromagnetic spectrum. While other pieces of the Big Bang picture fell into place, the missing matter remained a mystery.

There is an example much closer to home than the constellation Coma Berenices: the mass and distribution of observable matter in the Milky Way Galaxy isn't sufficient to account for the way the Galaxy rotates. What would it take to cause the Milky Way to rotate as it does? The matter should be mostly outside the visible disc of the Galaxy, it ought to extend well beyond the edge of the observable disc, and much of it should not be level with the disc but 'above' and 'below' it. If all that were the case, then the rotation would make sense. The suspicion is that the Galaxy must be surrounded by a halo of dark matter that is much larger than the observable mass of the Galaxy. The total diameter of the Galaxy might be four or five times what it is possible to observe in any range of the spectrum. Dark matter might also provide an explanation for the 'hat brim' tilt of the Galaxy's thin gas disc.

There is no way to investigate dark matter directly, only by watching how it affects other things – that is, what its gravitational effect is on other matter and radiation. Sometimes it gives its presence away by the manner in which it bends the paths of light. Such paths through spacetime are bent by the presence of massive objects ('benders') such as stars, planets, galaxies and galaxy clusters. This happens regardless of whether or not the benders are themselves detectable at any wavelength. When the distortion is too great to be caused by the observable matter in the bender, or when there is no observable bender at all, researchers know they are not observing everything that's out there between them and the background. They suspect the presence of dark matter.

The mystery of dark matter lies at the heart of the problem of measuring the age and the future of the universe. Calculating roughly whether there is sufficient observable matter in the universe to produce the gravitational attraction necessary to keep the universe at critical density, omega-equals-one, shows that the amount of matter observed directly with present technology falls far short. But the discussion doesn't end there, because dark matter does exist and because no one is yet certain how much there is or what it is.

Big Bang theory in its most straightforward form has it that even a microscopic deviation from omega-equals-one would have caused the universe very early on to recollapse or made it expand so rapidly that stars could never have formed. Inflation theory has proposed a solution to that fine-tuning problem, but the question right now is: *Is* the universe all that fine-tuned? It seems things would be dramatically different if it weren't, but it isn't actually obvious how it *is*. No measurement of existing mass density comes anywhere near critical density, which means that the universe should have expanded too fast for stars to form. It didn't. What is it we don't know yet?

Candidates for dark matter range from still-hypothetical mysterious exotic particles to black holes a billion times more massive than the sun. Primordial black holes (tiny ones formed in the early universe), planets, dwarf stars too dim to have been observed, massive cold gas clouds, comets and asteroids, and an assortment of dead or failed stars make up a broad middle ground of possibilities. Some physicists insist on tossing in a few copies of the *Astrophysical Journal*.

In 1998, hard-to-detect particles known as neutrinos moved to the short list. The existence of neutrinos is not a new idea. They were first suggested in 1930 by Wolfgang Pauli as a way to explain a mysterious loss of energy in some nuclear reactions, but it was not until 1956 that observations by Frederick Reines and Clyde Cowan at the Los Alamos National Laboratory in New Mexico confirmed their existence outside of theory.

No one now questions the existence of neutrinos, but they remain notoriously difficult to study. They rarely interact with any kind of matter. A typical neutrino can pass through a piece of lead a light year thick without hindrance. Clues to their existence come on those rare occasions when a neutrino does happen to collide with an atom, but even then the evidence is indirect.

Whether neutrinos have any mass at all has been in question, and of course if they have no mass they cannot be contributing to the mass density of the universe. There have been a number of claims in the last few years of the discovery of neutrino mass, but much stronger evidence came in June of 1998, from a team of Japanese and American physicists at an observatory in Takayama, Japan.

Their detector was a tank the size of a cathedral containing 12.5 million gallons of ultra-pure water, inside a deep zinc mine one mile inside a mountain. The rationale for the experiment ran like this: one of the ways neutrinos are produced is when cosmic ray particles from deep space slam into the Earth's upper atmosphere. Experimenters hoped to compare neutrinos that came from the upper atmosphere directly over the detector (a short distance) with those that were coming up from under the detector after having passed through the Earth (a long distance). Neutrinos from both sources, moving through the water, would occasionally collide with an atom. The result of such a collision is a scattering of debris, and the particles of that debris race through the water creating cone-shaped flashes of blue light called Cherenkov radiation. The light is recorded by 11,200 20-inch light amplifiers that line the inside of the tank. Researchers analyse the cones of light, finding the proportions of different sorts of neutrinos coming from each direction, and attempt to determine whether the neutrinos, which come in three types, change type on their journey from the upper atmosphere. If neutrinos can make this change, that means they must have mass.

Dr Yoji Tkotsuka, leader of the team and director of the Kamioka Neutrino Observatory, the site of the detector, announced that the evidence was strong for neutrino mass. 'We have investigated all other possible causes of the effects we have measured,' he reported, 'and only neutrino mass remains.'

Calculations based on these findings show neutrinos might (not everyone agrees they do) make up a significant part of the mass of the universe. Not that a single neutrino amounts to much. The mass of a neutrino turns out to be almost infinitesimal – about $1/500,000$ of the mass of an electron. But neutrinos nevertheless pack considerable clout by dint of their numbers, for there are about 300 of them in every teaspoonful of space. They outnumber other particles in the universe by a billion to one. In fact, the discovery of that tiny mass by the team in Takayama adds considerably to the mass density of the universe, by some calculations more than doubling it at a stroke.

As promising as this might appear, the discovery that neutrinos have mass can't account for all the missing matter that calculations show ought to exist. And though the combined mass of all those neutrinos may be enough to slow the expansion of the universe, it isn't likely to be enough to stop it or turn it around. The puzzle still has not been solved. The discovery of neutrino mass hasn't revealed the future of the universe.

In the mid-1980s, a panel of astronomers reviewing plans for use of the Hubble Space Telescope decided that the determination of an absolute distance scale outside the Galaxy and the discovery of the expansion rate of the universe – the Hubble constant – should be among the highest-priority projects undertaken by the telescope. A team of astronomers from the United States, Canada, Great Britain and Australia, led by Wendy Freedman of the Carnegie Observatories in Pasadena, California, received the largest allocation of time on the Hubble Space Telescope for a period of five years. The Extragalactic Distance Scale Key Project, as they call their work, involves trying to determine distances to nearby galaxies more accurately than ever before. These galaxy distances will then form the underlying basis for a number of other methods that can be applied at more remote distances, making possible several independent measurements of the Hubble constant.

Wendy Freedman is a native of Toronto, Canada, and one of her most vivid childhood memories is of a trip with her father to northern Canada, where they watched the stars and he explained to her how long it takes their light to reach us on the Earth. When Freedman entered the University of Toronto in 1975 she intended to study biophysics, but she soon switched to astronomy. She got her doctorate in astronomy and astrophysics from Toronto in 1984, then received a Carnegie Fellowship at the Carnegie Observatories, and in 1987 was the first woman to join the permanent faculty there, where she remains to this day.

At the heart of Freedman's Extragalactic Distance Scale Key Project lies the effort to measure Cepheid distances to 20 galaxies with the Hubble telescope. It is these distances that are then expected to provide an absolute scale for other methods which give only relative distances (Type Ia supernovae, Type II supernovae, the Tully-Fisher relation, and surface-brightness fluctuations – see pages 219–23).

In 1994, Freedman and her team were attempting to measure more precisely the distance to the centre of the Virgo supercluster. They found twenty Cepheids in the spiral galaxy M100 in the Virgo cluster, at the core of the supercluster, the first sure identification of Cepheids that far away. The Hubble data indicated that these Cepheids are approximately 56 million light years from earth. That was nearer than earlier estimates put the centre of the Virgo supercluster.

From this new distance measurement and M100's recession velocity (learned from its red shift), Freedman and her colleagues calculated a new value for the Hubble constant, about 80 kilometres per second per megaparsec. Experts led by Allan Sandage had previously calculated that its value was about 50, nowhere near 80. Thus began one of the most heated debates in modern astronomy – either an extremely significant controversy or much media hype about nothing, depending on which side of the issue you stand.

Freedman's announcement came as a shock. The Hubble findings were an embarrassment. Deciding between values of

50 and 80 was not mere nit-picking: if the universe is expanding so much more rapidly than previously thought, it follows that less time has elapsed since the Big Bang than the 10 to 20 billion years most experts had settled on. Depending on the density of matter in the universe, a Hubble constant of 80 means the universe must be only eight to twelve billion years old, probably nearer eight.

It isn't uncommon in science for new findings to challenge earlier thinking, sometimes eventually undermining what nearly everyone has been assuming was virtually unassailable scientific knowledge. But this challenge was one of the most disquieting so far in the 20th century, for astronomers were fairly certain, based on what they considered sound understanding of nuclear physics and the rate at which hydrogen converts to helium in stars, that some of the oldest stars in the Milky Way are 14 billion years old, probably even older. The universe can't be younger than the stars in it.

The glitch made newspaper front pages all over the world. Scientists ground their teeth. Since the future of astrophysics and astronomy depends on massive public spending, these branches of science have an enormous stake in maintaining their credibility. Researchers wonder what will motivate the allocation of funds for science now that the old Cold War rivalry is history. If public opinion is to favour continuing support for this wondrous but expensive adventure, it really doesn't do to have announcements indicating that tax money is buying nonsense! Like the Church in Galileo's day, the modern scientific establishment has much to lose if simple faith – in science this time round – is undermined.

Freedman and her team are young astronomers. Those whose numbers they were questioning are some of the most highly – and deservedly – respected older members of the astronomy community. Sandage had spent the best part of a lifetime developing new measuring techniques and making careful observations to arrive at the Hubble constant value of 50. But iconoclastic as the team's announcement was, it didn't come entirely out of the blue, nor was such a conundrum unprecedented. In 1929, Hubble himself calculated the Hubble constant to be 500 kilometres per second per megaparsec, making the universe younger than geologists knew the Earth was. Baade refigured H_o at 250. Sandage reduced it still further to 180, then to 75, then (in the mid-1970s) to 55 plus or minus 10 per cent. These corrections succeeded in making the universe old enough to allow for the formation of even the most ancient stars and globular clusters, but not before discussions had taken place that resembled those that now followed Freedman's announcement. Nor had Sandage and Tammann's value for the Hubble constant previously gone unchallenged.

In the late 1970s and 1980s, when Sandage had settled on a Hubble constant close to 50 and an age of the universe of 15 to 20 billion years, Gerard de Vaucouleurs of the University of Texas, for one, took serious issue with those numbers. Shortly before Freedman made her findings public in October 1994 there were other studies whose results implied that current estimates of the universe's expansion rate and age might be headed for another revision. A team led by Robert Kirshner of the Harvard-Smithsonian Center for Astrophysics, using the Cerro Tololo Inter-American Observatory in Chile, measured the expanding debris from five supernovae and judged the universe might be from 9 to 14 billion years old. But Freedman's team's calculations, based on data from the Hubble telescope, were more convincing than any of these other challenges to the older numbers.

Astronomers leave a wide margin for error in calculations like these, and the immediate temptation is to wonder whether the numbers are sufficiently fuzzy to allow the universe to be just barely old enough and the oldest stars just barely young enough. However, stars didn't pop into existence the instant the universe began. Estimating that stars are the *same* age as the universe would be unsatisfactory. There must be a cushion of at least a billion years after the beginning to leave comfortable time for them to form. The leeway in Freedman's numbers and in current estimates of the age of the oldest stars is not enough. For reference: a Hubble constant of around 50 indicates an age for the universe of around 15 billion years; a Hubble constant of around 70 or 80, a much younger universe – about 10 billion years or less.

A negative reaction to Freedman's team's announcement came almost immediately, and not unexpectedly, from Sandage, whose office was right down the hall from Freedman's at the Carnegie Observatories. Sandage had served under Hubble himself in this same observatory when it was known as Mount Wilson. According to Sandage, the glitch was being grossly overpublicized and its importance exaggerated – mostly media hype. There were plenty of possibilities of error in the Hubble team's results. In their measurements of the apparent magnitudes of the Cepheids, for instance, and in their assumption that the galaxy where these Cepheids are is actually in the centre of the Virgo supercluster. Perhaps instead it is in the foreground, nearer our own Galaxy. Arguing for that is the fact that M100 is a spiral galaxy, and it is elliptical galaxies, not spirals, that are more commonly found in the centres of clusters like Virgo. Freedman countered that her team had already taken that possibility into account in assigning a wide margin of error to their estimate. What's more, as they had also reported in their original paper, the relationship of Virgo to a more distant cluster, the Coma cluster, had made it possible to step out to there and calculate the Hubble constant at that distance – a calculation that gave the same result.

The question also arose whether the rate at which Virgo is moving away is a dependable indicator of the recession rate of the universe as a whole. Tammann reported that his studies indicated that Virgo is actually moving away more rapidly than the rest of the universe. Here, again, was the perennial difficulty of sorting out the actual 'Hubble flow' from all the other movement that's going on among and within clusters and superclusters. How to extract from this complicated picture the part of all that motion that is directly attributable to the expansion of the universe? Any sample of the universe is likely to give a faulty reading unless it is an extremely large sample indeed. No one knows for certain how large a sample that would have to be.

Freedman and her team hadn't claimed that their result settled once and for all the value of the Hubble constant, but neither were they convinced by the opposition. Sandage's own measurements had recently been challenged on another front. He had been using Type Ia supernovae for making his distance measurements. Some of the measurements that gave Sandage and his collaborators Hubble constants of around 50 were based on the assumption that these all reach the same maximum brightness and are good standard candles. However, Mark M. Phillips, an astronomer at the Cerro Tololo Inter-American Observatory in Chile, had recently found that not all Type Ia supernovae do have the same brightness characteristics. Brighter ones appeared to occur in spiral galaxies or galaxies with many bright stars. Phillips had developed a technique for analysing the light curves (how the supernova brightens and dims) to recognize these differences and make allowances for them, but, as of 1994, Sandage had not corrected his data. Ominously for Sandage, Robert Kirshner and his colleagues at the Harvard-Smithsonian Center for Astrophysics *had* corrected theirs, and their estimation for the Hubble constant had risen from 55 to around 67.

At the American Astronomical Society meeting in January 1995, just two months after Freedman's announcement, the new measurements and the controversy they'd stirred up about the value of the Hubble constant and the age of the universe were centre stage. Mark Phillips and Mario Humay, Phillips's colleague in Chile, were there reporting their measurement of 25 supernovae, some as far away as one billion light years.

Compensating for the differences in maximum brightness that Phillips had discovered, they'd arrived at a Hubble constant of 60 to 70, in the middle range between Sandage's measurements and those of Freedman's team. Freedman announced that the Hubble telescope had now measured distances to 40 more Cepheids in M100 and distances to two other galaxies in the Virgo cluster, M101 and NGC925. The new data were consistent with her team's earlier results.

Eight months passed, and in September 1995, Nial Tanvir at Cambridge University, with colleagues at Durham University in England and the Space Telescope Science Institute in Baltimore, estimated – based on fresh Hubble observations of Cepheids – a distance of 38 million light years to the M96 galaxy, in the direction of the constellation Leo. From this they inferred a distance to the much more remote Coma cluster. The team's calculation gave the universe an age of 9.5 billion years, give or take a billion.

In March of 1996, a year and a half after Freedman's initial announcement, Sandage and colleagues had rallied and were ready to report the results of their ongoing supernova study. Sandage, whistling into the wind, it seems, told an interviewer, 'We believe that this marks the end of the "Hubble wars".' In 1990 there had been a Type Ia supernova in the galaxy NGC4639. Sandage's team had been observing the light curve of this supernova since 1992. What was particularly significant about this investigation was that the Hubble telescope had been able to see individual stars in this same galaxy and 20 of them were Cepheids. From their brightness, Sandage's group had been able to calculate the distance of the NGC4639 galaxy as 82 million light years, and from that they knew the absolute magnitude of the Type Ia supernova in a way that didn't depend on comparing that supernova's brightness with the brightness of other Type Ia's. Applying this fresh knowledge to previous measurements of the apparent peak brightnesses of six other Type Ia supernovae, Sandage recalculated their distances and came up with a Hubble constant of 57, in the range he had been insisting on all along.

Sandage considered the case closed. Freedman and her colleagues were not won over. Sandage's new results, compelling as they seemed, didn't make the team's own Hubble findings go away or point up any flaw in their calculations. The Freedman team's numbers did come down a little. At the Princeton '250th Birthday Conference' in June 1996, she reported a value of 73, based on a combination of the distance measurements to Cepheids, the study of Type Ia and Type II supernovae, the Tully-Fisher relation and surface-brightness fluctuations. Freedman said that with so much data accumulating so rapidly, the debate might well be settled in the next three years. Indeed, in May 1999 she announced that the findings of her team at the completion of eight years of work indicated a Hubble constant value of 70 and an age for the universe of 12 billion years, with an uncertainty of 10 per cent. Sandage is sure that the debate will finally be settled close to his own number, but perhaps not until well into the next century, too late for him personally to enjoy his victory. He has said of de Vaucouleurs's death in 1995 (the de Vaucouleurs who was his critic), 'Very unfortunate. Anybody in the middle of a crisis should live to see the resolution of the crisis.'

What about the age of stars? There has been less controversy about that than there has been about the age of the universe. In the winter of 1996, study of some of the most distant galaxies ever observed showed them to be as much as 14 billion years old, shoring up faith in earlier calculations. However, in the late summer and fall of 1997, physicists at Case Western University in Ohio, led by Lawrence M. Krauss, re-examined the age of some of the oldest and most distant stars, using measurements from the Hipparcos satellite. They recalculated the age of globular clusters previously thought to be as much as 15 billion years old or even older. Hipparcos's measurements, made with unprecedented precision, revealed that these globular clusters are further away than earlier estimates had put them, and so, in order for them to appear as bright as they do, they must also be brighter than previously thought. And if they are brighter, that means they are burning faster and have evolved more quickly, making them younger – perhaps 11 billion rather than 15 billion years old.

Hipparcos cannot directly measure the distance to these globular clusters. They are in the Milky Way's halo, outside the Galaxy's main disc, too far away for parallax measurements even with Hipparcos. Instead, researchers used Hipparcos to measure the distance and brightness of other stars and compared them with stars of similar composition in globular clusters. Catherine Turon of the Paris-Meudon Observatory, who along with others has calculated 12.8 to 15.2 billion years for the age of a globular cluster known as M92, has admitted there are difficulties with such measurement: the stars used for comparison are often dim stars that have no metals or other heavy elements. Getting models adapted to such extreme objects with low metallicity is problematical. Processes such as fast rotation or metals sunk out of view into the star could skew the conclusions. Michael Perryman, a project scientist for Hipparcos, is even more sceptical. Hipparcos's own data have shown that some of the stellar models are spectacularly wrong. These new calculations of the age of stars did not put the age-of-the-universe paradox to rest.

However, also in the autumn of 1997, astronomers using the Hubble telescope to watch the collision of two galaxies called the Antennae observed at least 1,000 clusters of newborn stars forming from giant hydrogen clouds in the centre of the merging galaxies, indicating that globular clusters are not all so ancient. Some at least are emerging out of more recent galactic collisions.

Perhaps old assumptions about the history and ages of stars are not unshakeable after all. But Freedman points out that though not all globular clusters are as old as previously thought, a great many in our Galaxy *are*. Also, continuing studies of faint galaxies in the Hubble Deep Field argue for an older universe. They show that some elliptical galaxies were already well advanced in years at a red shift of 1.2. It doesn't help that *some* things are younger than previously thought.

Sandage, Freedman and their associates are treading on the frontiers of science – very recent and ongoing science. The Cepheids in the Virgo cluster that stirred up the controversy couldn't be seen at all before the Hubble Space Telescope, in fact not before its faulty optics were corrected in December of 1993, less than a year before Freedman's team's discovery. Some of the measuring techniques being used are barely past the experimental stage, yielding data whose implications no one fully understands. It would be the stuff of tabloids to declare anything settled. Those who are uncomfortable when science yields paradoxes rather than certainties will have to go on being uncomfortable for a while.

But suppose the Hubble constant does end up indicating that the universe is younger than some of its stars. It may seem relatively easy to think of the rapidity with which the universe is expanding as being the result only of a two-way tug-of-war between gravity (which is working to make the universe contract) and the expansion energy resulting from the Big Bang (which is working to make it expand). There may, or may not, be another player involved, and that player is Einstein's old 'mistake'.

Popular science books and articles habitually describe the cosmological constant as a repulsive force that might counter the effect of gravity. It would be well to acquire a little more sophistication and realize that the cosmological constant can actually work either way.

If the cosmological constant is a positive number, then it will indeed counter gravity, joining in the struggle on the side of expansion. However, if it is a negative number, then the effect will be to weigh in on the side of gravity. If it is zero, then it will do neither. Think of it then as a theoretical property of the vacuum of space that, if it exists and isn't zero, *might* act to stretch space and thus counteract gravity's contracting power, or *might* do the opposite. To put that another way, imagine yourself facing a dial. If the arrow is pointing to zero, then the cosmological constant is having no effect at all. If you move it to the minus side of the zero, the further you turn it the more it will contribute to the contraction of the universe. If you move it to the plus side of the zero, the further you turn it the more it will contribute to the expansion of the universe.

Even this slightly more sophisticated view of the cosmological constant does not begin to do justice to the complications involved when physicists and astrophysicists play with its value. The cosmological constant can seem to be working both ways at the same time, allowing us to have our cake and eat it too. However, what seems a contradiction is not, because of the way the cosmological constant fits into the equation for omega.

Theories of quantum mechanics – the study of the very small (atoms, molecules and particles) – have it that everywhere in the universe particles are spontaneously popping into and out of existence. Their life spans are unimaginably short. Nevertheless, 'empty space' seethes with this energy, and 'empty space' does not mean only what is out there dark and remote between the stars. This quantum energy fills the enormous amount of empty space within the atoms that make up chairs, tables, human bodies and all other things familiar and unfamiliar. 'Emptiness' is full of energy. Theory suggests that the energy of the cosmological constant might be this energy of virtual particles which wink in and out of existence at all times and everywhere in the universe.

For Einstein, the cosmological constant was only a mathematical device, and not long after he put it into his equations in order to avoid the implication that the universe must be either expanding or contracting, he decided it had been a mistake – for of course the universe *is* expanding. After visiting Hubble at Mount Wilson in 1931, Einstein rejected the whole idea, calling it 'theoretically unsatisfactory anyway'. But the cosmological constant didn't go away. Lemaître in particular enjoyed fooling around with it and adjusting its value, discovering that by fiddling with this theoretical dial he could construct universes that started out very slowly and then sped up, or universes that started out fast and then slowed down, or universes that began expanding, stopped and then expanded again. Something like that stop-and-start version was evoked in the 1940s as a possible remedy when new discoveries indicated that the universe was younger than the solar system. When it then turned out that the Hubble constant had been overestimated, the cosmological constant wasn't required after all and was packed away once again.

In 1948, researchers detected the effects, on atoms, of the vacuum energy decreed by quantum mechanics, but no one went on to study its possible influence on the universe as a whole until 19 years later when Zel'dovich, the Soviet theorist, realized that this vacuum energy would enter into Einstein's equations in just the same way that the old cosmological constant did. Before long it became evident that if Einstein had been right about mass causing spacetime to curve, and if all this vacuum energy really does exist, then the vacuum energy ought long ago to have curled the universe up into a tiny ball or something even smaller, or else driven the expansion so that even atoms – much less galaxies – could never have formed. Even by making the cosmological constant extremely small, Zel'dovich couldn't show how the universe could have turned out to be the way it is. So it seemed the value must be zero, and that is what most theorists since have been assuming. That zero does not, by the way, mean that there is no vacuum energy, only that by some truly remarkable coincidence, all the positives and negatives in that vacuum energy cancel out exactly.

As we've seen, the cosmological constant is still with us, hovering like a ghost in the equation for omega. If its value is zero we could write it off and forget it, but the symbol for it would still be sitting there. I am reminded of a recipe one of my more eccentric friends gave me for broccoli soup. The recipe had been passed from cook to cook, many times. Always the list of ingredients included a can of won ton soup, with parenthetical instructions: 'Don't put this in.'

Even before the recent discoveries that the universe may be expanding much faster than previously supposed, late-20th-century astrophysicists had been feeling once again an itch to reach for the cosmological constant dial. There was a possibility it might offer solutions to some intractable problems such as the still-missing dark matter. Nearly everyone was approaching the idea super-cautiously, having been burned twice before. John Noble Wilford commented in a *New York Times* article that one thing that makes physicists particularly uneasy about assigning the cosmological constant a value other than zero is that this reminds them too much of the way medieval astronomers designed increasingly complicated celestial mechanisms to explain the planets' motions, in order to preserve their beloved earth-centred Ptolemaic universe. As was true with those mechanisms, there was nothing to indicate that a cosmological constant value other than zero was wrong. But there was also nothing to indicate it was right. The best argument for it was that it allowed physicists to cling to theories in which they had a vested interest! Being able to turn the cosmological constant dial to a number (negative or positive) of their choice that ended up supporting the currently favoured version of the Big Bang theory was just a little too easy and allowed for too much leeway. With a friend like the cosmological constant, did a theory need enemies?

In 1990, Michael Turner of the University of Chicago and Fermi National Laboratory proposed a recipe to add up to critical density: 5 per cent ordinary matter; 25 per cent cold dark matter (including both invisible and 'exotic' types); 70 per cent the cosmological constant or something like it. According to Turner, the energy of the cosmological constant could compensate for some of the missing mass and serve as an additional brake on cosmic expansion, balancing things out in such a way that the universe would neither eventually collapse nor expand into an ever darker, thinner, colder infinity, but instead perch for all eternity on that highly desirable knife edge between the two. The lower density of matter that such a cosmological constant value would allow might be an added boon to theorists, making it easier to explain how matter congealed into such enormous structures as the Great Wall of galaxy clusters.

After Freedman's team's discoveries in late 1994, physicists began to consider much more seriously these suggestions that the cosmological constant might not be zero. Adjust the dial, and the cosmological constant's energy, over time, could change the rate at which the universe expanded. If expansion was slower when the universe was young, that would have given more time for stars and large structures to develop. Later, the energy of the cosmological constant could have influenced the expansion to speed up. The current measurements of the rate of expansion, by Freedman and others, would be only measurements of the *present* rate of expansion, and unreliable indicators of the age of the universe.

However, though a non-zero cosmological constant was looking more and more tempting in terms of explanatory power (in other words, useful to explain what was going on), there was still the major hitch that no one had yet been able to find any direct observational evidence that the value was anything other than zero. The first hint that this might change came in 1996, not, itself, from direct observation.

The age-old method of testing alternative ideas and making up for insufficient evidence by using mathematical simulations came into its heyday with the advent of supercomputers. In the 1980s and 1990s, as never before, it was possible to feed in some observed and assumed conditions in the early universe and find out what this might lead to after billions of years. In 1996, an international team headed by Carlos Frenk of Durham University in England, using Cray supercomputers at Munich and Edinburgh, ran simulations to find out whether temperature fluctuations observed in the early universe could have led from a Big Bang fireball, where everything was almost uniform, to today's universe of galaxies, clusters and voids.

The starting point for the simulations was the universe as it is thought to have existed 300,000 years after the Big Bang. That was when the cosmic microwave background radiation originated (the radiation that Penzias and Wilson detected in 1964). In 1992 George Smoot and his colleagues had been able to discern wrinkles, tiny energy fluctuations predicted by Big Bang theory, in this otherwise smooth cosmic fabric. Frenk's team simulated the growth of those initial wrinkles. The results supported the possibility that the cosmological constant should indeed be summoned from limbo.

Frenk and Simon White at Munich (assisted by Adrian Jenkins, Frazer Pearce and Joerg Colberg) ran four different simulations. One of the ways they differed from one another was in the estimates of the mass density of the universe. A second difference was in whether or not Frenk and his colleagues allowed the cosmological constant to be other than zero. Of the four (see Figure 8.2), the one capable of producing the universe as we know it today was the model in which the mass density was only 30 per cent of what experts think would be needed to produce omega-equals-one, or critical density, *and* in which Frenk also factored in a non-zero cosmological constant.

In this simulation, the expansion rate changed over time and was slower in the early universe than it is today. The computers demonstrated how the wrinkles might have attracted surrounding matter. Lumps of matter, according to the simulation, collapsed onto themselves and grew larger by merging with other lumps, eventually forming a complex filamentary network of large, twisting ridges surrounding vast empty regions. Gas and dark matter flowed along these filaments. Where the filaments intersected, galaxies and galaxy clusters formed. In the simulation, the last few billion years don't show much gross

Figure 8.2 Computer Simulations from Carlos S. Frenk and his colleagues

C most resembles the universe today. It is based on a mass density only 30 per cent of what we think would be needed to produce omega-equals-one, or critical density, and Frenk also factored in the cosmological constant.
A and **B** are much less successful models based on greater density.
D is based on 30 per cent density, without the cosmological constant.

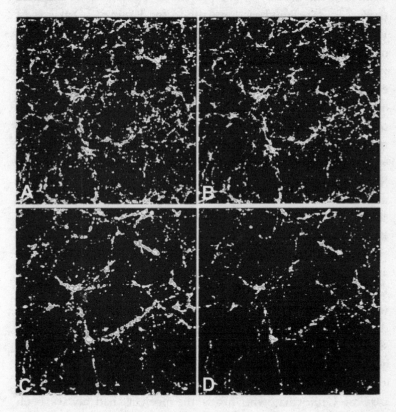

alteration, for the universe is expanding fast and the mass density is too low for the large structures to change very much.

No simulation, by itself, can provide the answers to the questions about expansion rate and age, the cosmological constant and the missing mass. But Frenk, in an interview with the *New York Times*, argued that his team's simulations do point out strengths and weaknesses in several theoretical models and 'give us greater confidence in what are, you might say, the best-buy models of the universe'. The project's results were in line with those of more modest simulations by Jeremiah Ostriker of Princeton and Paul Steinhardt of the University of Pennsylvania, and with models developed by James Peebles of Princeton.

The possibility raised by the simulations that omega does not equal one echoed some recent observational evidence. Studies of the spectra of galaxies in the X-ray range had been raising questions about proportions of ordinary matter and exotic dark matter. Also, the discovery of ever-larger galactic superclusters seemed unexplainable if omega does equal one. On the other hand, simulations by Joel Primack at the University of California, Santa Cruz, and scientists at New Mexico State University seemed to rule out the cosmological constant. 'No one,' said Peebles, 'should start collecting bets on a low-density universe.'

That was how things stood when, at the January 1998 meeting of the American Astronomical Society, the Supernova Cosmology Project – a team that had been studying supernovae to find out whether the expansion of the universe is slowing down – announced that not only does the expansion show no signs of slowing down, it actually appears to be speeding up. Their announcement was another blockbuster.

Saul Perlmutter of the Lawrence Berkeley National Laboratory in California, who heads the project, had always been deeply interested in the most fundamental questions of how the world works. As an undergraduate at Harvard and working towards a PhD at Berkeley, he'd become increasingly convinced that serving on teams involving hundreds of participants – as is common in modern world-class particle physics – would give a young physicist little chance to shape the research. How else to ask the fundamental questions? Perlmutter decided to try astrophysics, and that is where he is today, shaping research that may indeed lead to the answers to his questions. But his experience in fundamental physics has inclined him to be more patient and less resistant than some of the astronomy culture can be to projects that take years of single-minded pursuit to complete.

Ten years ago, Perlmutter began what promised to be a lengthy and difficult endeavour indeed – and at the outset something of a gamble – using distant supernovae as mile-markers to measure trends in cosmic expansion. When preliminary results satisfied him and others that supernovae could be used effectively for such measurement and that available technology should be up to the task, Perlmutter and his team dug in for a long-range investigation.

The most distant supernovae that Perlmutter's team had discovered by January 1998 were some seven billion light years away, meaning that by the time their light reached telescopes on Earth, seven billion years had passed since the stars exploded. By now that light is feeble, red-shifted by the expansion of the universe. The Supernova Cosmology Project involves comparing the light of these distant supernovae with the light of bright nearby supernovae to determine how far the faint supernova light has travelled. The distances combined with red shifts of the supernovae give the rate of expansion of the universe over its history, allowing researchers to determine how much the expansion rate may be speeding up or slowing down.

The remarkable predictability of Type Ia supernovae is what makes this project possible. Although all Type Ia supernovae don't have the same brightness, it turns out that their absolute luminosity *can* be learned by watching how quickly each supernova fades away. Type Ia supernovae in nearby galaxies are so predictable that the time the supernova explosion began can be determined just from a look at its spectrum, and the most distant supernovae also have precisely the right spectrum on the right day of the explosion. 'The real similarity of the details of these events,' says Perlmutter, 'can be seen in the beautiful spectra we get from the Keck telescope in Hawaii, the largest in the world.' Researchers breathed a sigh of relief when it was clear that Type Ia supernovae that exploded when the universe was half its present age behave essentially the same as supernovae do now, for this eliminated one worry about the reliability of the project's results – the question whether Type Ia supernovae have been different in different epochs.

Because the most distant supernova explosions appear so faint from Earth, happen at unpredictable times, and last for such a short while, the team performs a tightly choreographed sequence of observations, using telescopes around the world and the Hubble Space Telescope. Some team members survey distant galaxies using the largest telescope in the Andes Mountains of Chile, while others in Berkeley, California, receive that data over the Internet and analyse it to find supernova candidates. Once they find likely supernovae, they rush out to Hawaii to confirm that these *are* supernovae and measure their red shifts. Team members at telescopes outside Tucson, Arizona, and on the Canary Islands are meanwhile standing by to measure the same supernovae as they fade. The Hubble Space Telescope is summoned into action to study the most remote of the supernovae, whose distances make them too difficult to measure accurately from the ground.

By January 1998 Perlmutter's team had analysed 40 of the roughly 65 supernovae so far discovered by the project. Only a little earlier they had reported that the cosmic expansion rate seemed to have slowed down very little, if at all. Now Perlmutter was ready to report that 'all the indications from our observations of supernovae spanning a large range of distances are that we live in a universe that will expand forever. Apparently there isn't enough mass in the universe for its gravity to slow the expansion to a halt.'

In March 1998 a second research group reported similar findings. This team was headed by Brian Schmidt of the Mount Stromlo and Siding Spring Observatory in Australia and included Adam Reiss, a young astronomer at the University of California at Berkeley, and Kirshner from Harvard-Smithsonian. They reported that they had found indication that the expansion rate is approximately 15 per cent greater now than it was when the universe was half its current age.

No sooner were the words out of Perlmutter's and Schmidt's mouths than speculation began in earnest about what this news might mean for inflation theory and for the cosmological constant. Inflation theory predicts a flat universe. The new findings were indicating an open universe. Or were they? One extremely intriguing implication of the discovery was that these teams of astrophysicists might actually be looking at the first strong observational evidence that there is a repulsive force operating in the universe, that the universe is indeed getting an anti-gravity boost from somewhere. The evidence, said Perl-mutter, strongly suggested a cosmological constant.

No one was jumping to the conclusion that there are no other possible explanations. Michael Turner, who had proposed the recipe for critical density, reflected the caution of the scientific community when he said, 'If it's true, this is a remarkable discovery. It means that most of the universe is influenced by an abundance of some weird form of energy whose force is repulsive.' Schmidt said his own reaction was 'somewhere between amazement and horror. Amazement, because I just did not expect this result, and horror in knowing that it will likely be disbelieved by a majority of astronomers who, like myself, are extremely sceptical of the unexpected.' Reiss commented, 'We are trying not to rush to judgement on the cosmological constant. There could be some other sneaky little effect we have overlooked, something that makes the supernovae dimmer and appear to be farther away than they really are, or some variation in the behaviour of more distant supernovae that are deceiving us.'

In spite of such reservations, it seems Schmidt had overestimated his colleagues' scepticism, for by May a straw vote at a workshop at Fermi National Laboratory indicated that most scientists present agreed the two teams had made strong cases for an accelerating expansion rate and the existence of something resembling a cosmological constant.

It was all beginning to fit: the slowing down caused by the mass density of the universe appeared to be overwhelmed by the speeding up caused by the cosmological constant. Study of the relationship was telling researchers how much larger the energy density due to cosmological constant energy must be than the energy density due to mass density. Inflation theory predicts that omega equals one, and calculations showed that the cosmological constant energy could provide .75 of the total and the mass density .25. This proportion was close to the numbers coming from computer simulations and in Michael Turner's recipe. The discovery also held out hope for solving the age of the universe glitch, allowing the universe to have expanded more slowly at an earlier age.

But is the secret ingredient really that old ghost, the cosmological constant? Some have been calling it 'X-matter' and 'quintessence' (named after an element suggested by Aristotle) – speculative concepts in which textures in the early universe created conditions for a cosmic background energy. By the time of the workshop at FermiLab, cosmologists were referring to the 'missing energy' of the universe in the same way they had long spoken of the 'missing matter'. Some were calling it 'funny energy'. The mystery of what it is awaits early-21st-century physicists.

The Supernova Cosmology Project and Brian Schmidt's group hope to observe supernova observations even further back in time, to about 10 billion light years' distance. There are also proposals for studies involving new X-ray astronomy space-craft and for surveys of the cosmic microwave background radiation from the ground and from space. Clues to the density of the universe and the value of the cosmological constant are encoded in that radiation as the minuscule temperature variations that Smoot and his colleagues discovered.

Is it still necessary to ask about the deceleration parameter? There seems to be no deceleration! However, a discovery that the expansion rate is speeding up doesn't mean that the deceleration parameter must relinquish its place in the equation for omega. It can be either a positive or a negative number . . . and it can change over time.

With these new discoveries of the late 1990s, inflationary Big Bang theorists found themselves torn between glee and discouragement. One of the theory's greatest assets, its ability to solve the flatness problem, had become a potential embarrassment, for researchers were continuing to find insufficient matter to maintain a flat universe. The evidence that the expansion rate was speeding up could be taken as another nail in the coffin of a flat universe. If the universe was 'open', what good was a theory that predicted a flat universe? The theory had gone to a great deal of effort and brilliantly predicted a situation that might simply not exist.

However, speculation is rampant about the cosmological constant value, and whether 'quintessence' or 'funny energy' might in fact make up the deficit left by insufficient mass density, producing precisely the omega-equals-one flat universe that inflation theory predicts. Furthermore, since inflation theorists had already suggested that the cosmological constant might be the agent behind the inflation period in the early universe, observational evidence for its existence would be all to the good.

Another possibility for redeeming the theory is to reinterpret it to predict an open universe rather than a flat one, but most theorists balk at such a move. There is that danger –

reminiscent of Ptolemaic theory – of adding complications to a theory until it flounders not under its inability to explain and predict but because it can explain and predict too many contradictory findings.

This chapter has only barely sketched the problem of the elusive omega, giving a taste of the complications and the high hopes of modern researchers, and providing some background to explain announcements that will come in the next months and years. No one knows, at the moment, whether the arguments will continue for a long time with more and more disparate voices and conflicting data or whether there might actually be more definitive answers in the near future.

CHAPTER 9

Lost Horizons

When the universe was created, we were not consulted.

Andrei Linde

Whether there is an edge to the universe, and what, if anything, might be beyond, are old questions. German astronomer Heinrich Wilhelm Olbers, who lived from 1758 to 1840, pointed out what is now known as Olbers's paradox: if space has no edge and is infinite and contains an infinite number of stars, the night sky should be as bright as the Sun. It isn't. He wasn't the first to worry about that and certainly not the last. Suppose, instead, space is infinite but the number of the stars is not, and the stars are limited to some sort of system 'inside' infinite space. That creates another problem: their system will collapse because of their mutual gravitational attraction. Try to solve that by saying that the star system rotates and its centrifugal force keeps it from collapsing, and someone will surely think to ask: In relation to what is it rotating, since it's the only thing in an infinite universe?

An expanding universe takes care of some of those problems while introducing other challenges, particularly to non-expert thinkers: the balancing act summed up in the formula for omega, such proposals as Friedmann's first model in Figure 6.1 – a universe that is not infinite in size but nevertheless doesn't have any edges or boundaries in space (though it does in time), and the paradox of something that expands but doesn't expand 'into anything'. Theorists at the cutting edge of physics and astrophysics try to help us get our minds around these counter-intuitive concepts with stories of spheres and balloons, saddles and cones. We're as hard put as scholars were in the 16th and 17th centuries to decide how literally descriptions and the theories they represent are meant to be taken.

Clearly, all theories are not created equal when it comes to how much they should be accepted as 'reality'. ('Reality, whatever that may mean', quips Hawking.) For example, many physicists think that inflation theory is a description of something that very possibly happened – either that or something like it. Fewer are prepared to give wormhole theories or Hawking and Jim Hartle's no-boundary universe that much credence, but they do not reject them either. Then there is Hawking's most recent proposal, that the universe sprang into being from nothing, in the form of a particle of space and time resembling an extremely small, slightly irregular, wrinkled sphere in four dimensions – the 'pea instanton'. Though he may be right, you aren't required to salute that yet. None of the theories discussed in this chapter look likely ever to be confirmed from observational and experimental evidence in the way the Big Bang theory has been. Then again, who knows?

In a universe where on the large scale everything is moving further and further from everything else, if we reverse the direction of time and travel back towards the beginning, we will find things getting closer and closer together. In the late 1960s, Hawking and Penrose, taking off from Penrose's earlier work on black holes, showed that, in Hawking's words, 'if general relativity is correct, any reasonable model of the universe must start with a singularity' – that is, a point at which everything we will ever be able to observe in the universe was compressed in a point of infinite density. At the singularity, spacetime curvature would also have been infinite. A singularity is a dead end. Physical theories can't work with infinite numbers. All the theories of classical physics become useless there. No one can predict what would emerge from the singularity, and it's no use asking what happened before it. It's no wonder that when they find themselves locked out like this, physicists, whatever their religious or philosophical persuasion, are extremely ill at ease.

Hawking was disinclined to let the Big Bang singularity lie. He went on tinkering with his own ideas about what happens when things are very compressed, either in the centre of a black hole or in the very early universe. Eventually he and American physicist Hartle decided to employ a device called 'imaginary time' that theorists use for working out problems in quantum mechanics.

It isn't quite accurate to speak of Einstein's erasing the difference between the space dimensions and the time dimension, leaving four-dimensional spacetime. There remains a basic distinction in relativity theory between the space and time dimensions. That distinction disappears if the time coordinate is an 'imaginary number'. (See box on p. 286.) No longer are there three dimensions of space and one of time, or four dimensions of spacetime. In essence, there are four dimensions of *space*.

Hartle and Hawking's use of imaginary numbers and imaginary time is subtly different from the way others have employed them. They don't merely use this mathematical trick to solve a problem and then return to the more familiar concept of time. In their 'no-boundary' model, imaginary time is something that actually shapes the universe. They point out that imaginary time is a device used in quantum theory, and that is the theory dealing with the very small – atoms and elementary particles. In the early universe, everything was that small. The principles and concepts of quantum mechanics can be expected to apply there.

An **imaginary number** is a number that when squared yields a negative number. In everyday mathematics, the square of 4 is 16. The square of negative 4 is also 16. How can you square any number and arrive at negative 16? In the 17th century Gottfried Leibniz invented imaginary numbers to provide an answer to that question. The square of imaginary 4 is negative 16. The square root of negative 16 is imaginary 4.

Hartle and Hawking's suggestion is that if we were able to travel back towards what we have been assuming was the 'beginning' (the singularity) we would find, just short of reaching it, that (in imaginary time) it becomes meaningless to talk about 'past' at all. In a situation where there are four space dimensions and no time dimension, chronological time – with its well-defined past, present and future – would not exist, and with it would go all the vocabulary for describing chronological time. No more 'yesterday', or 'always', or 'past'. Discussions about a 'beginning' or 'before the beginning' would also have no meaning.

Hawking asks us to imagine travelling south on the face of the Earth. We can speak of ourselves as travelling south until we reach the South Pole, but there the concept of 'south' is meaningless. No one asks what is south of the South Pole. This is also a good analogy because there is no edge or boundary or beginning at the South Pole. Similarly there are no boundaries or edges or beginning in the Hartle-Hawking no-boundary universe – none in space and none in time. Does it follow that in this model time and space stretch to infinity? No. Just as is true of the surface of the Earth (it is finite in size), space and time are not infinite in the no-boundary universe. Hartle and Hawking's proposal doesn't compete with Big Bang theory, for in real time – the time in which we live – it would still appear to us that there was a singularity at the beginning of the universe.

Hawking refers to the idea as a proposal, not a theory. There are no direct observational data to support it. It is a wild but not illogical leap of imagination. In the 1980s and 1990s Hawking and others proceeded to ask what sort of universe would result from this no-boundary situation and to explore its connections with the observable universe of today. Needless to say, the calculations are extremely complex and so far they've been carried out only in simple models. However, until the late 1990s they seemed to indicate that there was no mathematical inconsistency between this proposal and the universe as we observe and experience it, or with well-accepted theories of modern physics.

Now comes a new challenge: the expansion rate seems to be speeding up. Could that happen in a no-boundary universe like Hartle and Hawking's? Or does it rule out their model? An accelerating expansion rate might mean the universe is 'open' and will go on expanding forever. A no-boundary universe that is analogous (though in more dimensions) to the shape of a sphere – like the Earth – *won't* go on expanding forever. It is a 'closed' universe, one that recollapses finally to a 'Big Crunch'.

But Hawking and Neil Turok, a colleague at Cambridge University, have thought of a way to look at the no-boundary universe as *either* spherical and closed and finite *or* open and infinite (shaped like the bell of a tuba). It depends on how you slice it. To demonstrate their new idea in fewer dimensions: it's like slicing a cone several ways. Cut through it horizontally, and the slice is a circle or a closed universe; vertically, and the slice is a hyperbola, an open universe. See Figure 9.1.

Inflation theory doesn't see the universe as self-contained in the way that Hartle and Hawking's no-boundary universe is. In fact, it makes suggestions about how our universe may relate to a much larger context.

Figure 9.1   Several Ways to Slice a Cone

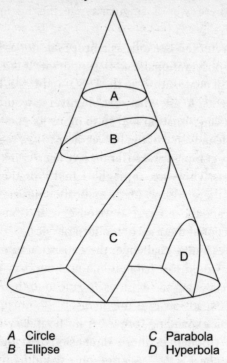

A Circle
B Ellipse

C Parabola
D Hyperbola

In order to demonstrate how inflation theory solved some of the problems in Big Bang theory, we inflated an imaginary balloon a little to represent the expansion of the universe before the inflationary period, paused to mark a tiny red dot on the surface of the balloon, and then inflated the balloon to a truly remarkable size. The tiny red dot itself became huge. It was the red dot, not the balloon, that represented the entire observable universe. Our 'universe' turned out to be only a very small fraction of everything there is. Draw a great number of red dots on the balloon, and when the gravitational repulsive force comes, each dot may respond differently. Some may not respond at all. Perhaps only one dot will expand. If so, that dot is our universe. What happened to the other dots? Are they universes too?

It is frankly unlikely that anyone will ever be able to discover whether our dot is unique. If the observable universe derived from a minuscule part of the initial conditions for the entire universe, it will be impossible to discover the ultimate extent and structure of Everything.

For anyone hoping that the quest to measure the universe will culminate with the revelation of the dimensions of everything there is, such news will inevitably be disappointing. Big Bang theory has it that the observable universe amounts to 75 to 90 per cent of the total universe. In the inflationary version of Big Bang theory, the observable portion is only a minuscule fraction of the total, and no one knows what fraction. If there were an infinite number of 'dots' on the 'balloon', even talking about a 'fraction' is incorrect.

Andre Linde has proposed something even more extensive – that each microscopic region that inflates is made up in turn of microscopic sub-regions, which inflate and are in turn made up of microscopic sub-regions – and so on and so forth – an eternal inflationary universe scheme. As Linde describes it, instead of being a single expanding fireball created in the Big Bang 'the universe is a huge, growing fractal. It consists of many inflating balls that produce new balls, which in turn produce more balls, ad infinitum.'

Is there any way to travel from one balloon or dot or ball to another? The idea of wormholes isn't new, nor is the notion (much utilized in science fiction films and television series) that they might offer a way of travelling to distant regions and times in the universe or to other universes. They were 'discovered' as a solution to Einstein's field equation in 1916 not long after he produced it. In the 1950s, American physicist John Archibald Wheeler led a research group that studied wormholes. Wheeler introduced the possibility of 'quantum wormholes'. It was these that captured the attention of Sidney Coleman of Harvard and Stephen Hawking in the 1980s. Coleman and Hawking took a particular interest in the possibility that such wormholes are part of the process by which new universes come into existence.

These quantum wormholes would be extremely small, only about 10^{-33} centimetres across. Written out as a fraction, that is 1 as the numerator and 1 followed by thirty-three zeros for the denominator. These tiny holes flicker into existence and then vanish after an interval too short to imagine. Again, think of an enormous balloon, the cosmic balloon, our universe. Picture dots on the balloon's surface. This time they represent not fledgling universes but stars and galaxies. Einstein predicted that massive objects curve spacetime, and the dots are doing that to the balloon's surface, causing tiny dimples and puckers. In spite of these, the surface is relatively smooth, even when examined through a microscope. It will take a more powerful microscope than any existing in our present technology to reveal that it is not smooth after all. What we are picturing is something no human being has ever seen except in theory or imagination, what the universe looks like on a scale 1,000 million million million times smaller than the scale of an atomic nucleus. The surface at this magnification is vibrating furiously, creating a frothy foam. At this level, fluctuations in the curvature of spacetime are not big, smooth curves like swells on the ocean. They are continuously changing ripples, crinkles and swirls. The 'surface' is hardly a 'surface' at all. It is like a bubble bath.

There are those, including Hawking, who like to quote the dictum of quantum theory that what is not forbidden can and will occur. It isn't surprising, then, to hear Hawking say that under high enough magnification the quantum fluctuation becomes such that there's a probability we'll find it doing 'anything'. The cosmic balloon might develop a minuscule bulge in it, and that in turn could become another tiny balloon, attached to the parent balloon by a narrow neck. The neck is a wormhole; the balloon is a baby universe.

It hardly needs saying that there are no experimental or observational data to support this speculative theory. Hawking is pessimistic about any tests revealing the existence of wormholes. No one expects ever to find direct observational evidence, for wormholes exist only in imaginary time, and even if that were not the case, their size rules out seeing them.

However, the newborn universe attached to this umbilical cord need not continue to exist only in imaginary time or stay small. If the theory is correct, our own presumably didn't. The new universe might end up something like ours, extending many billions of light years. Of course, with one universe spawning many more universes, and those many more yet, there must be a never-ending labyrinth of them. Successfully measuring the dimensions of our own universe would give us no clue what portion of the whole this represents.

These brief introductions to four models of the universe and beyond supply plenty of material for speculation, but the images described here do not allow us to say that the universe *is* or even *may be* shaped like a horn, or a sphere, or a pea, or a cone, or a parabola. These shapes are the closest theorists can come to representing in ordinary descriptive language what is only completely describable in the language of mathematics.

Einstein said that the gift of fantasy was essential to his work. Certainly these intellectual descendants of his who speculate about how the universe fits into a larger context have that gift. In one sense they are all remarkably inventive yarn-spinners. Yet their proposals are not science fiction. This is fantasy tethered to the known world by hefty guy-wires of mathematical equations. Nevertheless, no one knows whether these theories will be remembered only as curiosities, like Kepler's linking the planetary orbits to melodic lines, or whether like his laws of planetary motion they might turn out to be among the most significant advances in all the history of science. With present technology, it isn't even possible to make an educated guess which it will be.

EPILOGUE

Far and few, far and few
Are the lands where the Jumblies live.
Their heads are green, and their hands are blue,
And they went to sea in a sieve.

from 'The Jumblies' by Edward Lear

In the spring of 1997, one of the most breathtaking views of the Hale-Bopp comet was from the Lofoten Islands off the coast of Norway, north of the Arctic Circle. Fog often shrouds this remote, jagged land and seascape. But that spring when the mists cleared they revealed the comet suspended against a background of stars above an eerie billowing drapery of Northern Lights – all reflected in the water. Every now and then a streak of colour shot into the sky, soared in an arc over the comet, and disappeared among the stars.

Over the comet? Among the stars? So it might have seemed to those watching from the Lofoten Islands had they not known that the stars above them were much further away than the comet, not friendly sparkles but enormous, blazing infernos. The comet and its tail, that we could hide with our thumb held at arm's length . . . experts had told us that that was 50–60 million miles of light, dwarfing the earth. The Northern Lights are on our own doorstep by comparison and do not come anywhere near soaring in an arc 'over the comet'.

In these chapters we've followed human beings as they've pondered the magnificent enigma of that same sky for thousands of years with wonder, loneliness, and, as Galileo put it, 'with intense longing'. Men and women have written poems, sung about its beauty, worshipped it, found evidence there of a Creator God or no God at all, scrutinized it with telescopes, ventured a little way out. And, with remarkable success, they've measured the universe.

All tamed then? Is that how the story ends? The great mystery reduced to numbers and graphs?

My favourite analogy for the progress of science is one I learned from Herman Bondi: our scientific knowledge is an island – an island of 'What We Know' – that lies in the midst of a vast sea of the Unknown. As long as human beings have lived on the earth, they have been adding to that island, and the labour continues still at a frenzied pace. That just seems to be something humans *do*. They add to their island, as surely as squirrels gather nuts. So the island of What We Know grows larger and larger, spreading in all directions. Sometimes a cliff crumbles back into the sea, or we forfeit acreage in a hurricane, or a tidal wave sweeps away a promontory . . . and someone cautions Mr Elmendorf that a large portion of the land could shift alarmingly any day. But the island does grow inexorably, and it is astounding how much we now know! Unless the Sea of the Unknown is infinite, it surely must be shrinking.

Perhaps. But we notice something strange happening. Every time we add to the island of What We Know, the coastline – the line where we run up against the Unknown – grows longer. There are more and more locations where we encounter what we *don't* know.

In this book we've witnessed the living out of this parable

over the course of more than two millennia. We've seen evidence of it in the burgeoning cast of characters as the story progressed. In the early chapters, though these dealt with very long time spans, there were only a handful of men on the water's edge. We were able to enjoy leisurely biographical sketches of them. Later chapters discussed much shorter periods of time, but the number of persons involved grew. The biographies were shorter and more numerous. In chapters 7 and 8 it was barely possible to list all the names of those building on to the headlands, peering at the horizon, and putting out to sea in both small craft and exceedingly expensive contraptions.

The increase in sheer numbers is, of course, partly due to a growing world population and the availability of university-level education. The progression, as it manifests itself in this book, is also somewhat explained by the fact that the more hindsight we have, the better we're able to discern which were the significant insights and discoveries. Had we lived at the time of Copernicus, our discussion of contemporary astronomy would have had to include many more researchers, books and ideas than those mentioned in Chapter 2, and we might have chosen the wrong persons to profile. Four hundred years later it's easy to tell what work in the 16th century was leading to a dead end and what opened a door to the future, and to see that Copernicus dwarfed his contemporaries.

But the expanding number of people and projects also bears witness to the fact that the coast of our island is growing. There is more room on the shoreline every day, and people are eager to fill it. We're aware of so many more questions than our ancestors thought to ask, so many more areas that need investigation, so many more trails of evidence to be followed, anomalous details to make sense of, complexity to be unravelled, paradoxes to get our stubbornly intuitive minds around. Ptolemy, Copernicus, Galileo had no idea how mysterious this universe would turn out to be!

If the mystery isn't lost, perhaps the simplicity and single-mindedness of earlier science is? Has the coastline become too long? Too many names, too many teams, too much specialization, too many directions to go? Has it become nothing else but the exponential accumulation of arcane knowledge — more than anyone can ever fit together in a coherent picture?

Nature herself has made certain it isn't like that. We have learned that this reaching into the unknown, while it certainly encounters bewildering complications, always seems to grasp simplicity. The story appears to work backwards: at the end of the Hellenistic era, Ptolemy's explanation of the heavens was as complicated as the mechanics of Disneyland. Copernicus's description was a move towards simplicity. Kepler's more so. Newton's understanding was even more concise. Einstein's, simpler yet. Observations like those Galileo made with his telescope and those we are making with the Hubble telescope can be puzzling, can seem almost beyond possibility of explanation. Human genius searches for and often finds the beautiful symmetry that underlies and makes sense of the confusion.

We began this book by measuring a windmill, but nine chapters have shown us something subtly different from what my father, my brother and I did that day in Texas. We measured the windmill in an old-fashioned way, with hindsight, if you will. We knew that there were ways to measure it more directly, that its height was a number that could be found out precisely. We were not pushing our mathematics and technology to the limits by any means. The men and women in this saga of cosmic measurement *have* been doing that and more.

It is a particular fascination of mine to try to put myself in the place of those in earlier centuries who *didn't know what was coming next*. When we try to do this with the characters in this book, we are immediately reminded that at every stage of history, it was impossible for them to anticipate that what they could not do, their descendants *would* be able to do. Perhaps that's why they were so often willing to go out on a limb, to build on shaky assumptions, to fashion a precarious ladder, to put their faith in ill-understood calibrators . . . because they had no way of knowing whether it ever would be possible to stand on firmer ground.

Whatever the motivations, we have been an impatient and irrepressible lot . . . measuring the parallax of Mars in full knowledge that the uncertainty of the measurement would imperil the results . . . scrambling up the scaffolding to a giant telescope before it was safely braced . . . battling to measure stellar parallax long before the technological moment arrived . . . grasping at Cepheids as handholds into the universe without knowing the distance to even one of them . . . devising a formula for omega while our understanding of what goes into the formula is still worse than vague. In each epoch we have hoped that the impossible would later become easy. But we couldn't know, and we weren't willing to wait.

It might all have been less messy had we waited. It might in fact have more resembled what we would like to think the history of cosmic measurement has been. Many accounts give the impression that after Copernicus we 'knew' the arrangement of the solar system. After Cassini we 'had' the measurements to the planets and 'were sure of' the length of the base line afforded by Earth's orbit. In the 1950s we finally 'discovered' the size of the universe. It all moved by solid increments.

It hasn't been like that. We have groped, guessed, doubted one another, made missteps, built the rungs of the ladder too close together, felt it buckle beneath our feet, fought for a hold. In this book we've ended our tale not on the coastline of the island of What We Know but on jetties built far out from shore . . . even on small, frail craft almost out of sight of land . . . at sea in a sieve. But that is nothing new. Indeed that is precisely where we have been in every chapter of this book, at every stage of this history.

The nature we've striven to understand has fought back by showing us our place. Sometimes that's been a severe comeuppance, but it isn't all bad news. To be sure, human beings are not the centre of the universe, and they are not large by universal standards, but they aren't small either. Not the largest nor the smallest things around by a long shot. Draw a line from the smallest to the largest, and from our vantage point the ends of the line stretch in both directions to numbers beyond our ability to comprehend. As far as we can tell, we are somewhere near the midpoint of the line, approximately equidistant from the ends as we now perceive them.

There's another way to measure us. If we draw a line from the simplest to the most complex, we are not sitting at the midpoint. We're at one end. We are the most complex thing we have yet discovered in the universe. The human mind is still largely unexplained. The human situation is unfathomable. How paradoxical that with motives and longings and limitations rooted in the confusion of who we are, we probe the depths and heights, often with complex mathematics as our only tool . . . on a quest to discover not more complication, but simplicity!

'Who hath stretched a measuring line across it?' God taunts Job in the scriptures. Shall we raise a timid hand and venture, 'I think . . . well . . . actually . . . *we* have'? Maybe. Maybe not. For it is still a great mystery how large our island is − this treasured, hard-won, incalculably valuable, perhaps tiny island of human knowledge − compared with the sea.

INDEX